MW01223674

ELECTRICAL ESTIMATING

PROFESSIONAL REFERENCE

SECOND EDITION

Adam Ding

Published by:

www.DEWALT.com/guides

OTHER TITLES AVAILABLE

Trade Reference Series

- Blueprint Reading
- Construction
- Construction Estimating
- Construction Safety/OSHA
- Datacom
- Electric Motor
- Electrical – 2008 Code
- HVAC/R – Master Edition
- HVAC Estimating
- Lighting & Maintenance
- Plumbing
- Plumbing Estimating
- Residential Remodeling & Repair
- Security, Sound & Video
- Spanish/English Construction Dictionary – Illustrated
- Wiring Diagrams

Exam and Certification Series

- Building Contractor's Licensing Exam Guide
- Electrical Licensing Exam Guide
- HVAC Technician Certification Exam Guide
- Plumbing Licensing Exam Guide

Code Reference Series

- Building
- Electrical
- HVAC/R
- Plumbing

For a complete list of The DeWALT Professional Reference Series visit **www.dewalt.com/guides**.

This Book Belongs To:

Name: _____

Company: _____

Title: _____

Department: _____

Company Address: _____

_____

_____

Company Phone: _____

Home Phone: _____

DeWALT Electrical Estimating – 2nd Edition
Adam Ding

Vice President, Technology and
Trades Professional Business Unit: . . . . Gregory L. Clayton
Product Development Manager: . . . . Robert Person
Executive Marketing Manager: . . . . Taryn Zlatin
Marketing Manager: . . . . Marissa Maiella

For product information and technology assistance, contact us at **Professional Group Cengage Learning Customer & Sales Support, 1-800-354-9706.**
For permission to use material from this text or product, submit all requests online at **cengage.com/permissions**.
Further permissions questions can be e-mailed to
permissionrequest@cengage.com.

ISBN-13: 978-0-9797403-6-7
ISBN-10: 0-9797403-6-3

Delmar
5 Maxwell Drive
Clifton Park, NY 12065-2919
USA

Cengage Learning is a leading provider of customized learning solutions with office locations around the globe, including Singapore, the United Kingdom, Australia, Mexico, Brazil and Japan. Locate your local office at: **international.cengage.com/region**.

Preface

How sure are you about your numbers when signing a contract? Lots of people today learn by making mistakes, but this is dangerous. By improving the way you estimate jobs, you could increase your profit margin significantly.

"We know how to estimate. It's just that, when the job starts, we are short of labor." Does this sound familiar? It's time for a change. This book was written to introduce a clean approach towards estimating and bidding, and is filled with page after page of estimating tips, checklists, worksheets, and data tables. It walks you through each step of the process for obtaining opportunities, defining a complete scope, quantity take-off and pricing, proposal preparation and submission, contract negotiation, and change order procedures.

This is not just a book on how to count lighting fixtures or measure conduit lengths. It is aimed to help you get the job you want and make the money you want!

Best wishes,
Adam Ding

CONTENTS

CHAPTER 1 – *Estimating Preparation* 1-1

Decision to Bid. 1-1

Getting Bid Documents 1-3

Preparing a Preliminary Estimate 1-4

Figure 1.1 – Electrical Estimating Process 1-5

Figure 1.2 – What to Include on a Marketing Letter 1-6

Figure 1.3 – Factors to Consider in Bid Decision. 1-7

Figure 1.4 – Bid Load Schedule Worksheet 1-9

Figure 1.5 – General Job Information Worksheet. 1-10

Figure 1.6 – Electrical Design Documents

    Evaluation Worksheet. 1-11

    Electrical Drawings. 1-11

    Specifications. 1-13

    Conclusions 1-14

Figure 1.7 – Other Design Documents

    Evaluation Checklist 1-15

Figure 1.8 – Document Request Form 1-17

Figure 1.9 – Measuring Building Gross Floor Area. 1-18

Figure 1.10 – Taking Off Job Preliminary Information . . 1-19

Figure 1.11 – Estimating "Ball-Park" Numbers 1-20

CHAPTER 2 – *Quantity Take-Off*. 2-1

Defining Work Scope 2-1

Visiting Site. 2-4

Taking Off Quantities 2-4

Contacting Suppliers and Subcontractors. 2-7

Figure 2-1 – Electrical Scope Worksheet 2-9

    General. 2-9

    Equipment 2-10

    Distribution. 2-11

Motor Controls 2-12

Lighting. 2-13

Systems 2-13

Site Service 2-14

Equipment Hook-up. 2-15

Figure 2.2 – Request for Information Form 2-16

Figure 2.3 – Site Visit Worksheet 2-17

Figure 2.4 – General Quantity Take-off Checklist. 2-18

Figure 2.5 – Switchgear Quantity
Take-off Checklist. 2-19

Figure 2.6 – Lighting Fixtures Quantity
Take-off Checklist. 2-20

Figure 2.7 – Special Systems Quantity
Take-off Checklist. 2-21

Figure 2.8 – Miscellaneous Quantity
Take-off Checklist. 2-22

Figure 2.9 – Lighting Fixtures Quantity
Take-off Worksheet. 2-23

Figure 2.10 – Lamps Quantity Take-off Worksheet 2-24

Figure 2.11 – Lighting Fixtures Quantity
Take-off Tips 2-25

Figure 2.12 – Conduits Quantity Take-off
Worksheet. 2-26

Figure 2.13 – Fittings Quantity Take-off Worksheet 2-27

Figure 2.14 – Wired Quantity Take-off Worksheet 2-28

Figure 2.15 – Conduits/Wires Quantity
Take-off Tips 2-29

Figure 2.16 – Boxes and Devices Quantity
Take-off Worksheet. 2-30

Figure 2.17 – Boxes and Devices Quantity
Take-off Tips 2-31

Figure 2.18 – Equipment/Panels Quantity
Take-off Worksheet. 2-32

Figure 2.19 – Equipment/Panels Quantity
   Take-off Tips 2-33

CHAPTER 3 – *Pricing* **3-1**

Estimating Material Costs 3-2

   Do Your Own Take-off 3-2

   Get Quotes. 3-2

   Evaluate Quotes. 3-3

   Man-hours 3-4

   Labor Hourly Rate 3-6

Estimating Subcontractor Costs 3-7

Estimating Overhead Costs 3-8

   Jobsite Overhead. 3-8

   Office Overhead 3-9

Estimating Profit. 3-10

Getting a Correct Total 3-11

Figure 3.1 – Pricing Procedures 3-12

Figure 3.2 – Pricing Mechanics 3-13

Figure 3.3 – Material Pricing Worksheet. 3-14

Figure 3.4 – Supplier Evaluation Worksheet. 3-15

Figure 3.5 – Estimating Labor Hourly Rate. 3-16

Figure 3.6 – Labor Pricing Worksheet 3-18

Figure 3.7 – Salary Conversion Table. 3-19

Figure 3.8 – Subcontractor Evaluation Worksheet. 3-21

Figure 3.9 – Telephone Quotation Worksheet 3-22

Figure 3.10 – Electrical Jobsite Overhead Items 3-23

Figure 3.11 – Estimating Jobsite Overhead 3-24

Figure 3.12 – Estimating Office Overhead 3-25

Figure 3.13 – Estimating Profit Rate. 3-26

Figure 3.14 – Pricing Summary Worksheet 3-27

Figure 3.15 – Shortcuts for Estimating Mark-up 3-28

CHAPTER 4 – *Bidding* 4-1

Writing a Proposal 4-1

Cost Breakout Requirements 4-3

Assessing Risks 4-4

Submitting the Proposals 4-6

What To Do With Bid Shopping 4-7

Figure 4.1 – Bid Proposal Form 4-8

Figure 4.2 – General Proposal Scope Inclusion 4-9

Figure 4.3 – Distribution Proposal Scope Inclusion 4-10

Figure 4.4 – Wiring Proposal Scope Inclusion 4-11

Figure 4.5 – Fixtures Proposal Scope Inclusion 4-12

Figure 4.6 – Special Systems Proposal
    Scope Inclusion 4-13

Figure 4.7 – Proposal Scope Exclusions 4-14

Figure 4.8 – Cost Breakout Calculation 4-16

Figure 4.9 – Estimating "Risk-Dollars" 4-17

Figure 4.10 – Question List Before Sending a Bid 4-18

Figure 4.11 – Checklist for Telephone Bidding 4-19

Figure 4.12 – Question List for Renovation Jobs 4-20

CHAPTER 5 – *Post-Bid* 5-1

Post-bid Review 5-1

Proposal Follow-up 5-2

Improve Bid-hit Ratio 5-3

Estimating for Contract 5-4

Turn-over Meetings 5-6

Estimating for Project Management 5-7

Unit Pricing 5-8

Figure 5.1 – Common Bidding Mistakes 5-9

Figure 5.2 – Trips to Reduce Estimating Mistakes 5-10

Figure 5.3 – Post-bid Analysis Worksheet 5-11

Figure 5.4 – Post-bid Document Folder 5-12

Figure 5.5 – Question List for Bid Follow-up 5-12

Figure 5.6 – Common Value Engineering Ideas 5-13

Figure 5.7 – Bid History Tracking Worksheet 5-14

Figure 5.8 – Post-bid Mark-up Analysis 5-15

Figure 5.9 – Prepare a Revised Estimate
for Contract 5-16

Figure 5.10 – Elements in an Electrical Subcontract 5-17

Figure 5.11 – Example Work Scope in an
Electrical Subcontract 5-18

Figure 5.12 – Negative Clauses in a
Bad Electrical Subcontract. 5-19

Figure 5.13 – "Turn-over" Meeting Agenda 5-20

Figure 5.14 – Simplified Job Cost Codes. 5-21

Figure 5.15 – Change Order Pricing Procedures 5-22

Figure 5. 16 – Phone Memo Worksheet 5-23

Figure 5.17 – Unit Pricing Example 5-24

Figure 5.18 – Unit Pricing Worksheet. 5-26

Figure 5.19 – Estimating "Design-build" Jobs 5-28

Figure 5.20 – Electrical Contract Relationship 5-28

CHAPTER 6 – *Man-hour Tables* 6-1

Understanding Man-hour Tables 6-2

Adjusting Standard Man-hours 6-3

Writing Your Own Man-hour Tables 6-4

Daily Jobsite Man-hour Worksheet 6-5

Tips for Estimating and Improving Labor Productivity. 6-6

Converting Minutes to Decimal Hours 6-7

Lighting Fixtures. 6-8

Incandescent Light Fixture. 6-8

Track Light Fixture 6-8

Exit Light Fixture. 6-8

Fluorescent Light Fixture 6-9

T-bar Fluorescent Lighting Fixture 6-10

Discharge Lighting Fixture 6-11

HID Exterior Floodlights 6-11

HID Roadway Luminaries. 6-11

Ceiling Fans with Lights 6-11

Lighting Poles. 6-12

Lamps. 6-13

Outlet Boxes. 6-13

Switches. 6-15

Receptacles 6-15

Cover Plates. 6-15

Switchboards 6-16

Distribution Panels 6-16

Motor Control Centers 6-16

Circuit Breakers 6-17

Panelboards 6-18

Load Centers 6-19

Meters 6-19

Hubs. 6-19

Transformers – Dry Type. 6-20

Transformers Vibration Isolators. 6-21

Emergency Generator 6-21

Safety Switches – NEMA 1, 12, and 3R. 6-22

Safety Switches – NEMA 4 and 5. 6-23

Fuses 6-23

Relays. 6-23

Motor Starters 6-24

Signal Cabinets 6-25

Feeder Bus Ducts. 6-26

Plug-in Bus Ducts 6-27

Bus Duct Elbows 6-28

Bus Duct End Closures 6-28

Bus Duct Tap Boxes. 6-29

Bus Duct Circuit Breaker Adaptors 6-30

Time Clocks 6-30

Photo Cells. 6-30

GRS Conduits and Fittings. 6-31

IMC Conduits and Fittings 6-32

EMT Conduits and Fittings. 6-33

PVC Conduits and Fittings. 6-35

PVC Coated Steel Conduits and Fittings. 6-36

Aluminum Rigid Conduits and Fittings. 6-37

Flex Conduits and Fittings 6-38

Liquid-Tight Flex Conduits and Fittings 6-39

ENT Conduits and Fittings 6-40

Nipples 6-40

Service Head-clamp Type 6-41

Condulets. 6-42

Condulet Covers 6-43

Seal-off Fittings 6-43

Expansion Fittings 6-44

3-Piece Couplings 6-45

Bushings. 6-46

Insulated Bushings. 6-46

Locknuts. 6-47

Conduit Straps 6-47

Clamp Back 6-48

Enclosures 6-48

Wireway and Fittings 6-49

Underfloor Ducts and Fittings 6-50

Surface Metal Raceway and Fittings 6-52

Cable Tray and Fittings. 6-53

Copper Wires 6-54

Aluminum Wires – Types THHN, THW, XHHW, Etc. . . . 6-56

Flexible Cords 6-57

Low Voltage Cables 6-57

Building Wire Cables 6-58

Flat Conductor Cables and Parts. 6-59

Communication Cables 6-60

Residential Items/Devices 6-60

Residential (Romex) Cables 6-61

Equipment Hook-up. 6-62

Heating Devices 6-63

Signaling Systems 6-63

Pre-cast Concrete Pull Box/Manholes. 6-64

Pre-cast Concrete Transformer Slab 6-65

Motors 6-65

Trench Excavation 6-66

    Hand Excavation 6-66

    Machine Excavation. 6-67

Trench Backfill and Grading. 6-68

Pit Excavation. 6-69

Concrete Cutting 6-69

Channeling 6-70

Cutting Holes in Masonry. 6-71

Core Drilling 6-72

Sleeves. 6-72

CHAPTER 7 – *Estimating Forms* **7-1**

Figure 7.1 – Bid Load Schedule Worksheet 7-2

Figure 7.2 – General Job Information Worksheet. 7-3

Figure 7.3 – Building Area Worksheet 7-4

Figure 7.4 – General Quantity Take-off Worksheet. 7-5

Figure 7.5 – Lighting Fixtures Take-off Worksheet 7-6

Figure 7.6 – Lamps Take-off Worksheet. 7-7

Figure 7.7 – Conduit Take-off Worksheet. 7-8

Figure 7.8 – Fitting Take-off Worksheet 7-9

Figure 7.9 – Wire Take-off Worksheet. 7-10

Figure 7.10 – Boxes and Devices Take-off Worksheet. . . 7-11

Figure 7.11 – Equipment/Panels Take-off Worksheet 7-12

Figure 7.12 – Equipment/Panel Feeder
    Schedule Worksheet 7-13

Figure 7.13 – Material Pricing Worksheet. 7-14

Figure 7.14 – Supplier Evaluation Worksheet. 7-15

Figure 7.15 – Labor Hourly Rate Worksheet 7-16

Figure 7.16 – Labor Pricing Worksheet 7-17

Figure 7.17 – Subcontractor Evaluation Worksheet. 7-18

Figure 7.18 – Pricing Summary Worksheet 7-19

Figure 7.19 – Unit Pricing Forms for
    Residential Construction 7-20

    Duplex Receptacle. 7-20

    Ground-Fault Interrupter Receptacle 7-21

    20 Amp Outlet (Washer, Freezer, Etc.) 7-22

    Dryer Outlet 7-23

    Water Heater 7-24

    Air Conditioning Unit 7-25

    Air Handling Unit/Heater 7-26

    Range Outlets. 7-27

    Single-Pole Switch. 7-28

    Light Fixture Outlet. 7-29

    Well Pump 7-30

    Bath Exhaust Fan. 7-32

    Double Floodlight 7-33

    Telephone Outlet 7-34

    Door Chime and Push Button 7-35

    150 Amp Overhead Service 7-36

Unit Pricing Forms for Commercial Construction 7-38

    Duplex Receptacle 7-38

    Single-pole Switch 7-40

    Three-way Switch. 7-41

    2 × 4 Lay-in Fixture 7-42

    Phone Stub. 7-43

    3-Phase Rooftop Unit. 7-44

    200 Amp, 3-phase Overhead Service 7-46

    Emergency Light 7-48

    Bathroom Light. 7-49

CHAPTER 8 – *Technical Reference* **8-1**

Areas of Common Geometric Shapes 8-2

Volumes of Common Geometric Shapes. 8-3

Square, Cube, Square Root, and Cubic Root for
    Numbers from 1 to 100 8-4

Circle Circumference and Area (Diameters
    from 1 to 100). 8-7

Trigonometric Functions. 8-10

Master-Format 1995 Edition. 8-16

Master-Format 1995 Edition Electrical Titles 8-16

Master-Format 2004 Edition. 8-23

Master-Format 2004 Edition Electrical Titles 8-26

Uniformat Levels and Titles 8-32

Uniformat Electrical Titles – D 50 Electrical 8-33

Uniformat Electrical Titles – G 40 Site Electrical Utilities . . 8-35

Common Unit Conversions 8-36

Converting Inches to Decimals 8-51

Ohm's Law. 8-52

Common Electrical Quantities 8-52

Electrical Formulas. 8-53

Wire Sizes. 8-54

Wire Current Capacity 8-56
   Wire Types 8-56
   3 Wires in Cable – Ambient Temp. 86°F (30°C) 8-56
   Single Wire in Cable – Ambient Temp. 86°F (30°C) . . 8-57
Adjustments to Wire Current Capacity. 8-58
   Ambient Temperature Adjustment 8-58
   More Than 3 Wires in Cable Adjustment 8-58
Conduit Sizes. 8-59
Conduit Weight Table (Pound per 100 Feet, Empty) . . . 8-59
Supporting Spacing for Rigid Conduits 8-60
Copper/Aluminum Conduits and Wires 8-61
EMT, IMC, Rigid Conduit Fill Tables. 8-62
PVC Conduit Fill Tables 8-64
Flexible Metallic Conduit Fill Tables 8-65
Liquid-Tight Flexible Metallic Conduit Fill Tables. 8-66
Metallic Power Raceway Fill Tables 8-67
   500/700 Raceway. 8-67
   2000 Raceway 8-67
   2100 Raceway 8-67
   3000 Raceway 8-68
   4000 Raceway 8-68
   6000 Raceway 8-69
Non-Metallic Power Raceway Fill Tables 8-70
   400/800/2300 Raceway 8-70
   5000 Raceway 8-70
   NM 2000 Raceway. 8-70
Box Fill Tables 8-71
Motor Load Current 8-72
   Single-Phase Motor Full Load Current (Amperes) . . . 8-72
   Three-Phase Motor Full Load Current (Amperes) . . . 8-73
   Direct Current Motor Full Load Current (Amperes) . . 8-74

Transformer Load Current 8-75

   Single-Phase Transformer Full Load
      Current (Amperes) 8-75

   Three-Phase Transformer Full Load
      Current (Amperes) 8-76

Transformer Weight (lbs.) by kVA 8-77

Generator Weight (lbs.) by kW 8-78

Voltage Drop Table 8-79

Light Levels for Different Project Areas 8-80

Estimated Equipment Pad 8-81

Excavation Slopes 8-82

Estimating Trench Excavation 8-83

   For Excavation Slope of Vertical (or Angle
      Repose 90°). 8-83

   For Excavation Slope of 3/4:1 (or Angle
      Repose 53°). 8-83

   For Excavation Slope of 1:1 (or Angle
      Repose 45°). 8-84

   For Excavation Slope of 11/2:1 (or Angle
      Repose 34°). 8-84

Estimating Concrete for Conduit Encasement. 8-85

   Conduit Separation 1" 8-85

   Conduit Separation 11/2" 8-85

   Conduit Separation 2" 8-86

   Conduit Separation 3" 8-86

CHAPTER 9 – *Glossary* 9-1

CHAPTER 10 – *Abbreviations* 10-1

CHAPTER 1
Estimating Preparation

How much do you know about your numbers? How often is your estimate a guess? Estimating is where the profits are usually lost, even before the job starts. Prior to starting an estimate, you need to have a procedure in place. Figure 1.1 shows an overview of the estimating process for electrical contractors.

DECISION TO BID

If you are on a general contractor's bid list, you might receive notifications automatically for upcoming bids. The invitations could come in the form of fax, a phone call, e-mail or mail. If you have not heard anything from them in a while, it is better to give them a call to see if you are still on their bid list.

Other sources of business opportunities are:
1. Advertisement in a newspaper and trade journals
2. Bulletin posted in the offices of government agencies, school districts, state universities, private colleges, etc.
3. News services of construction trade associations and public plan rooms
4. Direct invitations from owners or architect/ engineers
5. Business contacts or word of mouth

DECISION TO BID *(cont.)*

Very often electrical contractors, especially new ones, have to team up with general contractors to win business, but why should they do business with you instead of others? You will have to convince them to put you on their bid list. When you do, be sure to provide more information than just a business card or letter of introduction. In Figure 1.2 there is a list of information you can offer for "pre-qualification".

Here are a few points worth mentioning:

1. If you decide to visit a general contractor's office, make an appointment beforehand.
2. Be honest with all the information you provide.
3. Ask questions so you can perform a background check on them.
4. Even if you do not appear to be a good fit for this contractor, ask for a referral for future opportunities.

Of course, you won't (and shouldn't) bid every job you are invited to. When deciding whether to bid a job or not, ask the essential question: *Is it a job you are likely to get and make a decent profit on?* If the answer is not likely, then you should walk away. Challenge yourself with the hard questions in Figure 1.3. Don't bother to bid on work that you don't have the tradesmen, money, or expertise to handle.

After deciding to bid on a job, put it in a bid load schedule as shown in Figure 1.4.

GETTING BID DOCUMENTS

Many electricians are used to requesting and waiting for drawings to arrive from general contractors. This wastes too much time and since only electrical sheets are expected, you may not receive the complete information. This may result in your bid being either too low or too high.

It is recommended that you visit the general contractor's office (or public plan rooms) where entire sets of documents are stored. Discuss the job with the general contractor's estimator to get a "general feel". Spend some time examining the documents and complete a job general information worksheet as shown in Figure 1.5.

Of course, you will need to go over electrical drawings/specs. An evaluation list is given in Figure 1.6 to determine the quality of the documents. It is also mandatory to check out the drawings for other trades. (See Figure 1.7 for another equally important checklist.)

At the end of the visit, ask the general contractor if they are willing to give you a copy of relevant drawings and specs (you may need more than the electrical portion). If so, fill out a document request worksheet (Figure 1.8) and give it to their estimator. Very often, you will have to buy them at your own expense from a reprographics company. In any case, make sure you have access to the complete information required for take-off and pricing.

PREPARING A PRELIMINARY ESTIMATE

Whenever possible, always prepare a detailed and accurate estimate. Before diving into the major details, it is helpful to have a general feel of what the "ball-park" number is.

Total building gross floor area (GFA) should be measured and compared to the architect's area number. Figure 1.9 offers some tips, while Figure 1.10 is a working example.

A preliminary estimate can be prepared with years of estimating experience and help of historical job cost data. Figure 1.11 provides two ways to do that:

1. Apply unit prices to the number of functional units (i.e. how many residential condos, hotel rooms, school students, hospital beds, etc.)
2. Apply unit prices to different functional areas (i.e. areas for parking, residential living, commercial, retail, or common facilities)

You might consider using both methods for cross-checking. The resulting number is the starting point of your estimate, but it should not be the ending point. It is helpful to get a rough idea of the total costs, which is to be verified later by more detailed estimates. A preliminary number should never be the only basis of your proposal since it is not accurate enough.

FIGURE 1.1 — ELECTRICAL ESTIMATING PROCESS

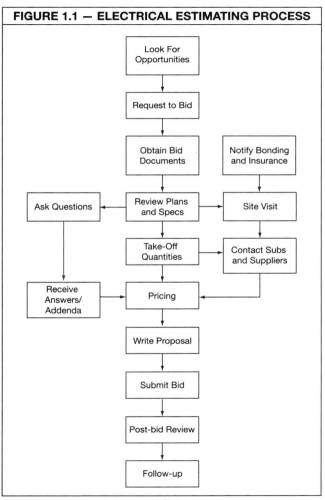

FIGURE 1.2 — WHAT TO INCLUDE ON A MARKETING LETTER

- Company name, address, telephone/fax number and e-mail addresses
- Is your company a corporation, partnership, or sole proprietorship?
- Years in business under the current company name
- Names of company officers, principals, partners, or owners
- The type and size of projects you would like to bid
- Type of work you do with your own forces
- Type of work you subcontract to others
- List of jurisdictions in which your firm is licensed
- Average annual dollar volume for the past three years
- Number of current employees in the office and field
- List of current jobs including contract amount
- List of completed projects including contract amount, completion date, architect name and general contractor name, contact and phone number
- Three or more business references including company name, contact person, and phone number
- Ability to bond and bonding capacity, bonding agent name, contact person, and phone number
- Insurance agent name, contact person and phone number, sample certificate of insurance
- Bank name and contact
- A list of trade associations of which you are a member

FIGURE 1.3 — FACTORS TO CONSIDER IN BID DECISION

- Is this job similar to the ones you normally do (i.e. type, size, complexity, quality, etc.)?
- Do you understand the scope of the work?
- Do you know the general contractor?
- Do they have a bad reputation with non-payments or bid shopping, etc.?
- Do they coordinate subs well?
- Is the general contractor using you to do a price check?
- Is this a real job or just a budgeting exercise?
- Are there many electricians bidding the job? Who are they?
- Have you bid against those guys before? Did you win?
- How far is this job from your office?
- Is the job in "your territory"? Are you licensed to do work over there?
- Do you know the material suppliers in that area as well as local electrical codes?
- Is the job in a heavy union area? Is your company a union or open shop?
- How is the owner's reputation? Do they have enough money to finance the job?
- How and when you will expect to receive the payments once the job starts?

FIGURE 1.3 — FACTORS TO CONSIDER IN BID DECISION (cont.)

- What are the procedures for change orders (if allowed)? How long do you have to wait before getting paid for changes?

- Who is the architect or engineer? How is their reputation?

- Are the drawings complete? Is the job still in the design phase?

- How many jobs do you have under way? If another job is awarded, will your crew be available to do it?

- Do you have to invest in special tools and equipment to do the work?

- Is there anything unusual about the job such as night shifts or limited working hours?

- Overall, can you expect a decent profit if you get the job?

FIGURE 1.4 — BID LOAD SCHEDULE WORKSHEET

This is an office schedule to control which job you are bidding every day.

Estimate #	Job Name	Job Size	Bid Date	Estimator
1012	XYZ Store	5,400 SF	03/04/05	AD
1013	ABC Shop	4,300 SF	03/10/05	BC
1014	Bay Condo	150 Units	03/15/05	AD
1015	MC Grocery	20,000 SF	03/18/05	AD
1016	John's Residence	1,800 SF	03/31/05	BC
1017	Auburn Hospital	200 Beds	04/01/05	AD
1018	Roadside Motel	30 Rooms	04/05/05	BC

FIGURE 1.5 — GENERAL JOB INFORMATION WORKSHEET

Job Name: __ABC Condo__

Estimate No. __1011__ Estimator: __AD__

Bid Due Date: __03/01/2005__ Time: __2:00 PM__

Job Location: __Any City, Any State__

Tax Rate: __9%__

Construction Type: __New Construction__

Gross Floor Area: __25,000 SF__ Number of Floors: __3__

No. of Condo Units: __20__

Send Bid To: __General Contractor__

Owner: __XYZ Development__ Tel. No: __(123) 456-7890__

General Contractor: __RST GC__ Tel. No: __(456) 789-0123__

Architect: __ACME Architecture__ Tel. No. __(789) 012-3456__

Electrical Engineer: __ACME Engineering__

Tel. No. __(789) 012-3457__

Job Start Date: __05/01/05__ Job Finish Date: __03/31/06__

Phases: __None__ Retainage: __10%__

Liquidated Damages: __$1,500/Day__

Warranty Period: __Standard__

Labor Conditions: __Open Shop__

Bond Required: __No__ Pricing Requirement: __Firm Price__

Send RFI to __General Contractor__

Work Done by Owner's Contractor: __None__

Cost Breakdown Requirements: __Building and Site__

Allowances: __Lighting Fixture $50,000__

Alternates: __None__

FIGURE 1.6 — ELECTRICAL DESIGN DOCUMENTS EVALUATION WORKSHEET

Job Name: _____

Estimate No.: _____   Estimator: _____

Electrical Drawings

___ Job physical address, type of occupancy, and any special conditions

___ Drawn to the same scale as architectural plans and printed full-size

___ Show names and uses of all rooms; north arrow and door swings

___ Separate from plans of other trades

___ Signed and sealed by electrical professional engineers

___ Consist of lighting, power, and auxiliary systems

___ Single line diagram showing

    ___ Service

    ___ Sub-panels and major loads

    ___ Breaker and fuse with sizes and types

    ___ Conduit and feeders with size and types

    ___ Grounding conductors and physical protection

    ___ New and existing devices

    ___ Service load calculations in amperes

___ Branch circuit diagram showing

    ___ Circuiting for lighting and general use receptacle outlets

FIGURE 1.6 — ELECTRICAL DESIGN DOCUMENTS EVALUATION WORKSHEET *(cont.)*

Job Name: _____

Estimate No.: _____ Estimator: _____

Electrical Drawings *(cont.)*

___ Conduit and wire to other circuits, equipment, motors, devices, etc.

___ Panel schedules showing

   ___ All panelboards and circuit loads

   ___ Number of lights and receptacles on each circuit

   ___ Total building connected load and calculated load demand in kilowatts

___ Light fixture and similar electrical equipment schedules with power data

___ Legends and details

___ Show all connections, permanent or temporary, inside and outside

___ Locations and sizes of all conduits, cables, and wiring

___ Circuits noted by numbers

___ Names and capacities of special outlets

___ Location and details of switchboards, motor control centers, power panels, lighting panels, and other equipment

___ Locations of fire alarm appliances and control panels

FIGURE 1.6 — ELECTRICAL DESIGN DOCUMENTS EVALUATION WORKSHEET *(cont.)*

Job Name: _____

Estimate No.: _____ Estimator: _____

Electrical Drawings *(cont.)*

___ Locations, connections and controls of signals, speakers, clocks, telephones, fire alarms, and other special systems

___ Indicate required code clearance areas of electrical equipment

___ Notes for special conditions (i.e. sloped ceilings, beam pockets, soffits, ceilings over 15 ft., hazardous or corrosive environments, elevated temperatures, etc.)

___ Show interlocks with other systems (i.e. fire sprinkler, HVAC, etc.)

Specifications

___ Basic electrical materials and installation methods

___ Manufacturers' names, products brands, or catalog numbers

___ Required performance criteria for all materials and assemblies

___ Coordination procedures, clean-up methods and inspections

___ Specialty work required

___ Equipment supplied and/or installed by others

___ Operations and maintenance manuals covering each item of equipment

FIGURE 1.6 — ELECTRICAL DESIGN DOCUMENTS EVALUATION WORKSHEET *(cont.)*

Job Name: _____

Estimate No.: _____ Estimator: _____

Conclusions

Do you think these documents are complete enough for bidding?

❏ Yes ❏ No

Are drawings and specifications coordinated?

❏ Yes ❏ No

Are electrical drawings coordinated with other trades?

❏ Yes ❏ No

Are documents in compliance with governing electrical codes?

❏ Yes ❏ No

Issues to be resolved with the design engineer:

_____

_____

_____

_____

FIGURE 1.7 — OTHER DESIGN DOCUMENTS EVALUATION CHECKLIST

Civil/Landscaping Drawings

- Where the wiring begins: property line, at the structure, or 10 ft. from the building?
- If a transformer is required, who's paying and who installs it?
- Any site lighting required for parking lot?
- Any site features to require power, like fountains?

Architectural Drawings

- How many floors and what is the square footage for each floor?
- Is the architect's number for the building area correct? Is he measuring from inside corner of the building? Is he counting balconies in or out?
- What is the distance between floors?
- What is the ceiling height for each floor?
- Where are electrical rooms on each floor?
- What does the interior designer require about lighting fixtures? Is there a schedule on interior design drawings?

Structural Drawings

- What type of construction for each floor?
- Is the lowest floor concrete slab on grade or wood framing with a crawl space?
- Are upper floors wood framing, steel joist/decking, concrete suspended slab, or hollow-core pre-cast concrete?
- How the structural details are going to affect conduit runs and fixture installation?

FIGURE 1.7 — OTHER DESIGN DOCUMENTS EVALUATION CHECKLIST *(cont.)*

Mechanical Drawings

- What type of electrical service does mechanical equipment require?
- Is mechanical contractor to provide wiring for controls, motors, disconnect switches, motor starters, etc.?

Specifications
(Division 1 General Requirements)

- Instruction to bidders
- Owner, architect, and engineer
- Chain of command and responsibility
- Bonding and insurance requirements
- Supplementary and special conditions
- Completion dates and liquidated damages
- Payment terms, schedule, and hold back
- Change order procedures
- Dispute resolution
- Cash allowances
- Alternates
- Maintenance, turn-over, and training

FIGURE 1.8 — DOCUMENT REQUEST FORM

Job Name: _____

Estimate No.: _____ Estimator: _____

To (General Contractor/Owner's Rep/Blueprint Shop):

_____

We need the following documents in order to bid the above project:

Drawings	Sheets/Sections
Specifications	
Electrical	
HVAC	
Plumbing	
Fire Protection	
Civil	
Landscape/Irrigation	
Architectural	
Interior Design	
Structural	

Please deliver the above documents to:

Contact Person: _____

Company Name: _____

Delivery Address: _____

Telephone Number: _____

Thank You.

1-17

FIGURE 1.9 — MEASURING BUILDING GROSS FLOOR AREA

Measuring Procedure

1. Measure areas from exterior corners of building perimeter wall.

2. Find out how many individual floors the project has (basement, main floor, upper floors, loft, penthouse, etc.). Don't forget upper stories.

3. Subdivide each floor into smaller segments that are easier to measure. Break down into shapes: rectangles, squares, triangles, circles, and semicircles.

4. Calculate each shape separately and add them to get the area for each floor.

5. Add up each floor to get the total building gross area.

6. Don't include areas that carry no wiring (i.e. covered walkways, canopies, patios, balconies etc.)

FIGURE 1.10 — TAKING OFF JOB PRELIMINARY INFORMATION

TAKING OFF GROSS FLOOR AREA

Floor	Area (Sq. Ft.)	Perimeter (Ft.)	Floor Height (Ft.)	Ceiling Height (Ft.)
Parking	1064	169	11.0	9.0
Level 1	9515	527	11.5	8.0
Level 2	9256	426	9.0	8.0
M/E Level	1626	184	9.3	8.0
Total Area	21461	Sq. Ft.	—	—

TAKING OFF FLOOR STRUCTURE AREA

Floor	SOG (Sq. Ft.)	Suspended Concrete (Sq. Ft.)	Wood Framing (Sq. Ft.)
Parking	1064	—	—
Level 1	7758	1757	—
Level 2	—	—	9256
M/E Level	—	—	1626
Total Area	8822	1757	10882

TAKING OFF UNITS (I.E. CONDO UNITS)

Units	One-Bedroom (Ea.)	Two-Bedroom (Ea.)	Three-Bedroom (Ea.)
Level 1	14	1	1
Level 2	14	1	—
Total	28	2	1

FIGURE 1.11 — ESTIMATING "BALL-PARK" NUMBERS

Note: All numbers and dollar amounts in these examples are fictional and only for illustrative purposes.

By Function Unit:

Electrical costs for a hotel of 110 rooms:
$110 \times \$6,000 = \$660,000$

By Function Area:

In a high-rise condominium:

Parking/Retail Area
60,000 Sq. Ft. $\times \$2.75 = \$165,000$

Residential Suites Area
120,000 Sq. Ft. $\times \$8.70 = \$1,044,000$

Total Electrical Costs
$\$165,000 + \$1,044,000 = \$1,209,000$

CHAPTER 2
Quantity Take-Off

A good or bad estimate begins with the take-off. You need to get use to a set of routine procedures and apply them to every job estimate. By doing so, nothing is left out and you will save yourself a lot of work in the long run.

DEFINING WORK SCOPE

You shouldn't always immediately start estimating by taking off quantities. Spending adequate time reviewing documents can reduce errors and omissions. Remember, better slow than sorry.

1. Check Documents
Once you receive a set of bid documents, open the package and check to see if the required information is there. For example, electrical specifications may have been left out or some drawing sheets could be missing.

If the package seems to be complete, then flip through the pages to see what qualities these documents have. Make sure they are readable and appear to be construction documents, not schematic designs.

2. Study Drawings

Thoroughly study drawings, sheet by sheet, until they are fully understood. Go over each floor, trace panels, and conduits. Visualize how components fit into the job and develop an overall mental picture. Mark special notes on plans with different colored pencils (i.e. blue for switchgear, green for fixtures, pink for devices, etc.).

It is suggested that you take out all drawing sheets that show schedules, riser diagrams, installation notes, etc., and hang them on the wall in your office. That way, you can cross-reference drawings easily.

3. Read Specifications

Read through the specifications, paragraph by paragraph, at least twice. Specs can be complex, consisting of a whole division with many pages of detailed information, or they can be simply laid out on the same page as the drawings. In any case, missing an "oddball" note in the specs can be disastrous!

Make notes in the following aspects of the specifications:

- Scope of work (what is to be included)
- Types of conduits allowed for different applications
- Expensive and special items
- Manufacturer, type, and size of equipment
- Work not shown on drawings, but required
- Work to be coordinated with other trades

• Warranty, service period, instructions, test, and balance, etc.

4. Deal with Conflicts

After reviewing the drawings and specifications individually, you can verify if the two are compatible. If they conflict with each other, normally specifications shall overrule. However, when drawings and/or specs conflict with the governing electrical code, most of the time the code will overrule.

Every electrical estimator should be familiar with the National Electrical Code (NEC®), but some cities and counties have special requirements. Find out which version of NEC® applies to your job and also stay current on local electrical codes.

If you do not feel comfortable with any issue, you should get clarification from the design engineer.

5. Summarize Scope

Use the worksheet in Figure 2.1 to make notes. Write down as much as you can, including types and dimensions of equipment, conduit sizes and material, etc. Make reference to related drawing sheets and specification sections. The completed worksheet is an ideal template to define the job scope.

For unclear details, you'd better ask the engineer rather than making assumptions. Problems must be called to the attention of the architect/engineer early enough to allow time for issuing an addendum.

DEFINING WORK SCOPE *(cont.)*

Use an RFI form (Figure 2.2) to send questions. If you do not get an answer after some days, follow up with a phone call. Architects should issue an addendum including answers from the electrical engineer. Through addenda, your competitors will also include additional costs in their bids.

VISITING SITE

Before you bid a job, you always need to visit the site. Very often, site conditions are different from what you see on electrical drawings. A site visit will enable you to make additional allowances to account for items that are hard to quantify.

You can use the Site Visit Worksheet in Figure 2.3 as a general guide. *Note:* sometimes repeated site visits might be necessary to get more information. It is recommended to start each estimate within 24-hours of site inspection.

TAKING OFF QUANTITIES

To "take-off" a job means that you "take" the information "off" the documents and translate it into a list of items with quantities.

1. Take-off Breakout

It is important to decide how you should break out your take-off. Sometimes owners have their detailed cost breakdown requirements which you must follow.

Otherwise, depending on each job, you can break out your take-off by systems, floors, phases, base price and alternates, etc.

2. Take-off Sequence

Lots of estimators follow this sequence:

- Lighting fixtures and lamps
- Wiring devices
- Conduits, connectors, and wire
- Boxes and covers
- Service and distribution (one-line diagram)
- Special systems
- Site electrical
- Equipment hookup
- Miscellaneous material and installation

3. Take-off Process

The actual take-off can be described in three easy steps:

A. Measure Each Item

General guidelines are as follows:

1. Do not scale drawings. Instead, try to use dimensions specified.
2. Each system should be taken off separately. You can apply color coding to differentiate. For example, conduits can be taken off in three main categories: power distribution, branch power, and branch lighting.

3. Try to visualize the way the work should be done. For example, do not forget to account for vertical risers when measuring conduits.

4. Remember to include required items that may not appear on the drawings including: fittings, hangers, shields, inserts, bolts, nuts, gaskets, etc.

5. Identify major conduit runs and quickly trace them out.

6. Note the number of similar spaces. Repetitive details like the same types of condo units on a residential job can save time in an actual quantity take-off.

7. Beware of reduced size drawings, as the scale noted on them is almost always not to be used.

8. Use a calculator that prints on paper (cashier type).

Because you might not be able to complete a quantity take-off without interruptions, mark drawings to remind you of what has been done.

Figure 2.4 through Figure 2.8 provide some electrical quantity take-off checklists for your reference.

B. Record Quantities

Because there is so much information you take off the drawings, it is better to use some sort of form to get organized. Good estimating forms provide a permanent record and reduce the chances of error. Figure 2.9 through Figure 2.19 list a few completed quantity take-off forms, as well as estimating tips.

While using these forms, it is important to keep different items separate. For example, floor outlets are different from wall outlets. Also, please make detailed references when recording data. Note which sheet you find items on (drawing and detail number, grid reference, etc.); where the items exist in the building (floor level, room number, etc.); and what details these items have. For example, the word "panel" without any other description is not clear when there are several panels in this job.

C. Summarize Quantities

Take-off quantities one drawing sheet at a time. When you are done, calculate the total quantities at the bottom of each worksheet. Round up the numbers, according to the precision desired (i.e. you might just want conduit lengths to be whole numbers without decimals). Then add up the quantities on each sheet and group them for the same type of items. The result can be later used for material quote requests and pricing.

CONTACTING SUPPLIERS AND SUBCONTRACTORS

Normally suppliers are to be contacted for material quotes on lighting fixtures and switchgear equipment. In addition to suppliers, electrical work often requires the use of subcontractors. The reason is, there are so many new specialties that no electrician can be knowledgeable in all of them. Normally, system manufacturers can refer you to their dealers in your local area, who can do the work for you.

CONTACTING SUPPLIERS AND
SUBCONTRACTORS *(cont.)*

After you decide to bid a job, immediately contact the subsystem specialty contractors in order to get their quotations, which might include:

- Data and communications
- Fire alarms
- Lightning protection
- Sound system
- CCTV
- Clock
- Intercom
- Security
- Nurse call
- Doctor's register
- Electrical heat

Depending on the scope, you might also need quotes from suppliers and subcontractors including concrete, painting, site electrical utilities, trenching and excavation, etc.

In your take-off, first count systems that require quotations, then notify the related suppliers and subcontractors as early as possible. Go over the job with them and provide relevant information (maybe an extra copy of drawings and specs). Without timely notification, subs either can not finish the bid or just bid high to make sure their costs are covered. This will raise your total bid which could cause you to lose the deal.

FIGURE 2.1 — ELECTRICAL SCOPE WORKSHEET

Job Name: _____

Estimate No.: _____ Estimator: _____

General

____ Underground
    Location: _____   Size: _____
    Excavation: _____   Source: _____

____ Equipment
    Recessed Depth: _____   Size limitation: _____ ___
    Method of feed: _____   Supports: _____ ___

____ Distribution
    Outlet locations: _____   Wiring materials: _____
    Method of feed: _____   Chases: _____
    In walls: _____ Under floor: _____ Overhead: _____

____ Fixtures and Devices
    Location: _____   Support: _____
    Finish: _____ Color: _____ Material: _____

____ Work by Specialty Subs
    Material Storage: _____   Coordination: _____

____ Scheduling
    Work to be done: _____ How long: _____
    By whom: _____
    Work to be done: _____ How long: _____
    By whom: _____
    Work to be done: _____ How long: _____
    By whom: _____
    Work to be done: _____ How long: _____
    By whom: _____

FIGURE 2.1 — ELECTRICAL SCOPE WORKSHEET (cont.)

Job Name: _____

Estimate No.: _____ Estimator: _____

Equipment

____ Switchboard

New: _____ Existing: _____ Revised: _____

Voltage: _____ Volts; Main Current: _____ Amps

Main Switch/CB: _____ Type: _____ Size: _____

Fused: _____ Size: _____

No. of Sections: _____

Interrupting Capacity: _____

Ground Fault Protection: _____

____ Transformers

Size: _____ kVA: _____

Type: _____ Dry: _____ Oil: _____ Liquid Filled: _____

Fan Cooling: _____

____ Unit Sub-station

Primary Voltage: _____ KV

Type of Incoming Line Section: _____

Type of Transformer: _____

Secondary Voltage: _____Volts

Type of Outgoing Section: _____

Indoor: _____ Outdoor: _____

Interrupting Capacity: _____

____ Standby Generator

New: _____ Existing: _____

Size: _____ KW Voltage: _____Volts

Manufacturer: _____ Type of Fuel: _____

FIGURE 2.1 — ELECTRICAL SCOPE WORKSHEET *(cont.)*

Job Name: _____

Estimate No.: _____   Estimator: _____

Equipment *(cont.)*

    Transfer Switch: _____ Type: _____ Size: _____

    Exhaust System: _____

    Indoor: _____ Outdoor: _____ ____

____ UPS

    Voltage: _____ Static: _____ Rotary: _____

    Capacity: _____ Battery Duration: _____

    Maintenance By-Pass: _____

    Load Bank: _____

____ Other Equipment: _____

Distribution

____ Distribution Feeders

    Voltage: _____ Volts; Type: _____; Feeding: _____

____ Bus Duct

    Type: _____ Copper/Aluminum: _____

    Capacity: _____ Amps

    Poles: _____ Neutral: _____ Feeding: _____

____ Secondary Feeders

    Voltage: _____ Volts; Type: _____;

    Transformer: _____ Volts

____ Power Panels

    Type: _____ Capacity: _____ Amps _____

    Circuit Breakers: _____ Fused: _____

FIGURE 2.1 — ELECTRICAL SCOPE WORKSHEET (cont.)

Job Name: _____

Estimate No.: _____ Estimator: _____

Distribution (cont.)

    Interrupting Capacity: _____

    Copper/Aluminum Bus: _____

_____ Lighting Panels

    Voltage: _____Volts; Type: _____

    Interrupting Capacity: _____

    Copper/Aluminum Bus: _____

_____ Grounding

    System Ground: _____ Isolated Ground:_____

    Building Ground: _____

_____ Cable Trays

    Type: _____; Material: _____

Motor Controls

_____ Motor Controllers

    MCC: _____ Class: _____ Starters: _____

    No. of Sections: _____ Fused: _____

    Breakers: _____

    Power Factor Correction: _____

_____ Motor Wiring

    No. of Motors: _____ Voltage: _____ Volts

    Total H.P.: _____

    Type of Air Conditioning: _____

    A/C Voltage: _____ Volts

FIGURE 2.1 — ELECTRICAL SCOPE WORKSHEET *(cont.)*

Job Name: _____

Estimate No.: _____ Estimator: _____

Lighting

____ Lighting Fixtures

   Supplied by Owner: _____ Allowance: _____

   Voltage: _____ Volts;

   Type of Lamps/Ballast: _____

   Main Fixture: _____ Type of Lens/Louver: _____

   Exterior Wall Fixtures: ____ Special Lighting: ____

____ Lighting Control

   Switching: _____ Dimming: _____

   Occupancy Sensors: _____

   Quick Connect Wiring: _____

   Control Consoles: _____ Relay Panels: _____

____ Branch Wiring: _____

Systems

____ Fire Alarm

   New/Existing: _____ Manufacturer: _____

____ Type: _____

____ Communications

   Telephone: _____ PBX: _____

   Enterphone: _____ Intercom: _____

   Computer/Data: _____ Radio: _____

____ Security

   Security: ____ CCTV: ____ Exterior Service: _____

FIGURE 2.1 — ELECTRICAL SCOPE WORKSHEET (cont.)

Job Name: _____

Estimate No.: _____ Estimator: _____

Systems (cont.)

____ PA System

New/Existing: _____ Manufacturer: _____

Type: _____

____ Electrical Heating

Type: _____ Voltage: _____ Controls: _____

____ Other Systems

Lightning Protection:_____

Lab Wiring: _____

Nurse Call: _____

Doctor Paging:_____

Staff Register:_____

Entertainment TV: _____

Snow Melting: _____

Underfloor Duct: _____

Site Service

New: _____ Existing: _____ Revised: _____

Voltage:_____Volts; Capacity: _____ Amps

Interrupting Capacity: _____

Cable Size: _____ Type: _____

Overhead: _____

Power/Tel. Companies: _____

Connections: _____

Excavation/Trenching: _____

FIGURE 2.1 — ELECTRICAL SCOPE WORKSHEET (cont.)

Job Name: _____

Estimate No.: _____ Estimator: _____

Site Service (cont.)

Concrete Encasement: _____

Equipment Pad: _____

Area Lighting: _____ Street: _____

Parking lot: _____ Walkway: _____

Type of Poles: _____ Pole Bases: _____

Irrigation Power: _____ Exterior Signage: _____

Access Box: _____ Pre-cast Concrete: _____

Cast Iron: _____ Flat Steel:_____

Equipment Hook-up

HVAC Units: _____

Elevator: _____

Doors (electrical locking, overhead, automatic entrance, etc.): _____

Kitchen Equipment:_____

Other Equipment: _____

FIGURE 2.2 — REQUEST FOR INFORMATION FORM

REQUEST FOR INFORMATION

FROM:	RFI #:
	DATE:
	JOB NAME:
	JOB REF. NO.:

TO:

ATTN:

We are requesting that you review the following matter and advise on how we are to proceed. In the event that your determination constitutes a change to the contract documents, please issue an addendum if possible.

ISSUE:

Please do not hesitate to contact us if you have any questions.

SIGNATURE:

ENCLOSURES:

REPLY:

SIGNATURE:	DATE:

FIGURE 2.3 — SITE VISIT WORKSHEET

Job Name: _____

Estimate No.: _____ Estimator: _____

Date: _____ Address: _____

Directions:_____ Distance from Home Office: _____

Site Access: _____

Contact Name: _____ Phone: _____

Competitions Present at Site Meeting: _____

Climate Conditions: _____ Rain/Snow/Wind: _____ Temp: __

Accommodations: _____ Costs Per Day: _____

Site Description: _____ Hazardous Material:_____

Site Office Location: _____ Laydown: _____

Parking: _____ Material Unloading: _____

Security: _____

Site Power Available: _____

Amps: _____ Phase: _____ Voltage: _____

Points of Connections for Utilities: _____

Pole/Panel/Hook-up/Energy Cost:

By Owner/GC: _____ By Us: _____

Telephone: _____ Water: _____ Sanitary: _____

By Owner/GC: _____ By Us: _____

Existing Building Maintenance Conditions:

_____

Demolition/Relocation/Upgrade of Existing Electrical System:

_____

Neighborhood Information: _____ Local Electrical Codes: ____

Local Supplier: _____ Labor Conditions: _____

Equipment Required: _____

Dust Control/Noise Abatement Requirements: _____

Comments: _____ Photos Attached: _____

Specific Questions: _____

Answers: _____

FIGURE 2.4 — GENERAL QUANTITY TAKE-OFF CHECKLIST

- ❏ Electrical permit
- ❏ Mobilization and demobilization
- ❏ Service set-up and utility company charges
- ❏ Temporary power and lighting
- ❏ Insurance and bond
- ❏ Demolition of existing electrical system
- ❏ Mechanical control wiring and starters
- ❏ Forklifts, cranes and special hoisting
- ❏ Equipment and tools
- ❏ Access panels
- ❏ Cutting and patching
- ❏ Roof penetration
- ❏ Fixture support
- ❏ Panel backing
- ❏ Scaffolding
- ❏ Testing
- ❏ Painting
- ❏ Dust protection
- ❏ Firestopping and smoke seal
- ❏ Sleeves
- ❏ Disconnect switches
- ❏ Interlock wiring
- ❏ Primary transformer
- ❏ Primary raceway
- ❏ Primary cable
- ❏ Job clean-up
- ❏ Material freight and sales tax
- ❏ Supervision and nonproductive labor
- ❏ Labor to install owner furnished items
- ❏ Travel expenses
- ❏ Office overhead
- ❏ Cash allowances
- ❏ Contingency
- ❏ Profit

FIGURE 2.5 — SWITCHGEAR QUANTITY TAKE-OFF CHECKLIST

- ❏ Main switchboards
- ❏ Distribution panelboards
- ❏ Load centers
- ❏ Motor control centers
- ❏ Motor starters
- ❏ Power panels
- ❏ Lighting panels
- ❏ Transformers
- ❏ Terminal cabinets
- ❏ Feeder cables
- ❏ Circuit breakers
- ❏ Fuses
- ❏ Meters and meter centers
- ❏ Capacitors
- ❏ Safety switches
- ❏ Manual starters
- ❏ Lighting contactors
- ❏ Bus ducts
- ❏ Time clocks
- ❏ Push-button stations

FIGURE 2.6 — LIGHTING FIXTURES QUANTITY TAKE-OFF CHECKLIST

- ❏ Fixtures
- ❏ Lamps (incandescent, fluorescent, mercury vapor, etc.)
- ❏ Plaster frames
- ❏ Remote ballasts
- ❏ Hangers
- ❏ Stems
- ❏ Spacers
- ❏ Couplings
- ❏ Tandem units
- ❏ Suspension systems
- ❏ Floodlight poles
- ❏ Brackets
- ❏ Valance and cove lighting
- ❏ Special lenses (i.e. plastic or glass lenses, parabolic louvers)
- ❏ Egg crate and reflectors
- ❏ Signs
- ❏ Security lighting

FIGURE 2.7 — SPECIAL SYSTEMS QUANTITY TAKE-OFF CHECKLIST

❑ Site lighting

❑ Fire alarm and signaling

❑ Telephone, intercom, cable, and data

❑ Audio/video

❑ Security and CCTV

❑ Underfloor duct

❑ Clocks and clock systems

❑ Hospital (nurse call, doctor's register, etc.)

❑ Lightning protection

❑ Electric heat, snow melting, and heat tracing

❑ Emergency power and UPS

❑ Energy management

❑ Substations

❑ Overhead transmission

FIGURE 2.8 — MISCELLANEOUS QUANTITY TAKE-OFF CHECKLIST

❑ Conduit and Wires
- Conduits (all types: rigid, EMT, flexible, PVC, etc.)
- Underfloor and wall ducts
- Cable tray
- Conduit fittings (elbows, locknuts, bushings, straps, clamps, hangers, condulets, expansion fittings, nipples, etc.)
- Enclosures and cabinets
- Terminals and lugs
- Wire and cable (all types)
- Grounding

❑ Outlet boxes and plaster rings

❑ Wiring devices
- Switches, dimmers, and sensors
- Receptacle (all types)
- Finish plates (all types)
- Special outlets
- Time switches and photo cells

❑ Equipment hook-up

❑ Trenching, excavation, concrete, core drilling, and light bases

❑ Demolition

FIGURE 2.9 — LIGHTING FIXTURES QUANTITY TAKE-OFF WORKSHEET

Job Name: __ABC School__

Estimate No.: __901__   Estimator: __AD__

Date: __Jan. 1, 2008__   Worksheet Page Number: __1__

Fixture Type	Basement	1st Floor	2nd Floor	Subtotal
A	3	1	2	6
B	—	2	—	2
C	—	—	4	4
D	4	5	6	15
E	2	1	—	3
F	—	1	—	1
G	3	—	—	3
H	—	—	3	3
I	—	2	—	2
J	—	—	1	1
K	1	—	—	1
L	—	1	—	1
M	—	—	1	1
N	—	2	—	2
O	2	—	1	3
P	1	1	—	2
Q	—	2	4	6

FIGURE 2.10 — LAMPS QUANTITY TAKE-OFF WORKSHEET

Job Name: __ABC School__

Estimate No.: __901__   Estimator: __AD__

Date: __Jan. 1, 2008__   Worksheet Page Number: __2__

Fixture Type	Subtotal	Lamp		
		F-40 CW	F-40	HPS
A	6	12	—	—
B	2	—	8	—
C	4	4	—	—
D	15	—	—	15
E	3	—	3	—
F	1	2	—	—
G	3	—	6	—
H	3	—	—	3
I	2	—	—	4
J	1	3	—	—
K	1	—	1	—
L	1	1	—	—
M	1	—	—	1
N	2	—	4	—
O	3	3	—	—
P	2	—	4	—
Q	6	—	—	6
TOTAL	—	25	26	29

2-24

FIGURE 2.11 — LIGHTING FIXTURES QUANTITY TAKE-OFF TIPS

1. Review the installation details for each fixture; make notes about ceiling conditions and mounting heights.

2. Count all the special components that will be required: hangers, spacers, caps, safety clips, wires or cables, guards, etc.

3. Pay attention to special fixtures: with radio suppressors, remote ballasts or diffusers, installed in coving, etc. Those are more expensive.

4. Identify the voltage and switching controls and look for dimmers.

5. Find out from suppliers whether lamps are included with the fixture. If not, you need to add material and labor costs for lamps. If yes, most likely you still need to add the installation labor.

6. To estimate the total quantity of lamps, multiply the number of lamps indicated for each fixture by the total number of fixtures, and then totals each type of lamp.

FIGURE 2.12 — CONDUITS QUANTITY TAKE-OFF WORKSHEET

Job Name: **ABC School**

Estimate No.: **901**   Estimator: **AD**

Date: **Jan. 1, 2008**   Worksheet Page Number: **3**

Conduit Type/Size	Basement	1st Flr.	2nd Flr.	Subtotal
1/2" EMT (2 #12)	200	420	550	**1,170**
1/2" EMT (3 #12)	90	130	290	**510**
3/4" EMT (3 #12)	30	—	20	**50**
3/4" EMT (4 #10)	20	—	80	**100**
1" EMT (3 #8)	3	—	5	**8**
1 1/4" EMT (3 #3)	30	—	—	**30**

FIGURE 2.13 — FITTINGS QUANTITY TAKE-OFF WORKSHEET

Job Name: __ABC School__

Estimate No.: __901__   Estimator: __AD__

Date: __Jan. 1, 2008__   Worksheet Page Number: __4__

Fitting Type/Size	Basement	1st Flr.	2nd Flr.	Subtotal
1/2" Connector	171	68	124	**363**
3/4" Connector	30	—	23	**53**
1" Connector	4	—	5	**9**
11/4" Connector	2	—	—	**2**
11/4" Elbow	2	—	—	**2**

FIGURE 2.14 — WIRES QUANTITY TAKE-OFF WORKSHEET

Job Name: **ABC School**

Estimate No.: **901**   Estimator: **AD**

Date: **Jan. 1, 2008**   Worksheet Page Number: **5**

Conduit Type/Size	Conduit Length	WireSize			
		#12	#10	#8	#3
1/2" EMT (2 #12)	1,170	2,340	—	—	—
1/2" EMT (3 #12)	510	1,530	—	—	—
3/4" EMT (3 #12)	50	150	—	—	—
3/4" EMT (4 #10)	100	—	400	—	—
1" EMT (3 #8)	8	—	—	24	—
1 1/2" EMT (3 #3)	30	—	—	—	90
TOTAL	—	4,020	400	24	90

FIGURE 2.15 — CONDUITS/WIRES QUANTITY TAKE-OFF TIPS

1. Find out what types of conduits are allowed or not allowed.

2. Find out allowable sizes for each conduit type, minimum and maximum.

3. Measure each type of conduit separately. Use colored pencil to mark different conduit types on the drawings.

4. Measure each size of conduit separately, working from the smallest size to the largest, within the same conduit type.

5. Check the plan scale before you start measuring conduits.

6. Don't forget to measure the vertical conduits running down on the wall.

7. Allow one coupling for each 10 feet of conduit and one for each elbow.

8. Calculate the wire needed from the conduit lengths. But watch for changes in the wire size on long runs. Sometimes a larger wire size is needed in the first portion of a run to reduce the voltage drop at the end of the line.

9. Allow extra wires at each outlet box and panel.

10. When taking off the underground conduit, start a separate worksheet for trenching, excavation, surface cutting, concrete or sand encasement, etc.

11. Make notes of special situations such as short-run conduits, exposed conduits on finish surface, through concrete or masonry walls, etc.

12. Apply waste only after you are done with all take-off (5% is normally enough based on good measurement).

FIGURE 2.16 — BOXES AND DEVICES
QUANTITY TAKE-OFF WORKSHEET

Job Name: __ABC School__

Estimate No.: __901__   Estimator: __AD__

Date: __Jan. 1, 2008__   Worksheet Page Number: __6__

Devices/ Boxes	Basement	1st Floor	2nd Floor	3rd Floor	Subtotal
Switch Box	1	5	5	5	**16**
S.P. Switch	1	5	5	5	**16**
Switch Plate	1	5	5	5	**16**
4" Oct. Box	10	130	130	120	**390**
4" Oct. Cover	10	130	130	120	**390**
4" Sq. Box	6	8	4	4	**22**
4" Plaster Ring	6	8	4	4	**22**

2-30

FIGURE 2.17 — BOXES AND DEVICES QUANTITY TAKE-OFF TIPS

Box and Rings:

1. It is best to actually count boxes and plaster rings one by one as shown on the plans instead of making allowances for their quantities.

2. Make notes regarding the size and type of each box. Consider the max number of conductors. Even if the specs allow a smaller size, you may have to use a larger box to comply with the code.

3. Examine the architectural details at each box to determine the plaster ring size. For example, if the wall has 1/2" thick drywall over 1/4" plywood, use a 3/4" deep ring.

4. If in a hurry, allow one outlet box for each single fixture and one for each row of fixtures.

Wiring Devices

1. Pay attention to special devices made for particular purposes, including those for heavy use with high ampacity and voltage ratings, in specification grade, with custom colors, with automatic grounding etc., as they cost more.

2. Watch for specs that require one type of device at some areas and another type at other areas.

3. Count switches first, then the receptacles and plates.

4. Three-way switches come in pairs. When you find one, look for the other across the room, down the hall, or up a flight of stairs.

FIGURE 2.18 — EQUIPMENT/PANELS QUANTITY TAKE-OFF WORKSHEET

Job Name: __ABC School__

Estimate No.: __901__   Estimator: __AD__

Date: __Jan. 1, 2008__   Worksheet Page Number: __7__

Equipment/Panel	Description	Quantity
SWITCHBOARD		
Main Switch	1200 A	1
C/T Compartment	—	1
Transition Section	—	1
Distribution Section	—	3
Fusible Switch	3P, 100 A	6
	3P, 200 A	4
	3P, 400 A	1
PANELBOARD WITH BREAKERS		
225 A, m.lug, 42 cir, 120/208	—	2
225 A, m.lug, 42 cir, 277/480	—	4
100 A, m.c.b., 24 cir, 120/208	—	1
SAFETY SWITCH		
H.D. Fused, 240 V, 3P, 200 A	—	2
H.D. Fused, 600 V, 3P, 60 A	—	3
TRANSFORMER		
45 kVA, 3P, 120/208	—	3

FIGURE 2.19 — EQUIPMENT/PANELS QUANTITY TAKE-OFF TIPS

1. Begin estimating equipment when you're fully familiar with the project (maybe the last one to take-off).

2. Make sure the quote from the supplier covers all equipment and panels.

3. Estimating checklist for equipment/panels

   __ Location __ Installation access __ Indoor __ Outdoor

   __ Service voltage __ Interrupt capacities

   __ Under voltage protection

   __ Size __ Fit in space __ Physical clearance per code

   __ Single-phase __ Three-phase

   __ Required connection __Overhead __Underground

   __ Special grounding __ Ground fault provisions

   __ Standby power equipment __ Special metering equip.

   __ Fuses __ Shunt trip breakers

   __ Surge arrestors __Transfer switches

   __ Unloading __ Special rigging

   __ Lifts/scaffolding __ Uncrating

   __ Flush mounted __ Surface mounted

   __ Housekeeping pads __ Equipment pads

   __ Anchors __ Leveling __ Final assembly __Testing

   __ Name plate __ Corrosion-resisting finish

   __ Temporary protection

   __ Utility company hook-up charge

NOTES

CHAPTER 3
Pricing

Pricing is the process in which you convert the quantities you took off into dollar values. Never guess, for what seems to be inexpensive may be very expensive. No matter how large or small a job, identifying every cost item is the key in predicting an accurate price.

Figure 3.1 lists general steps to do pricing, while Figure 3.2 shows graphically the mechanics in a pricing process.

An important principle to keep in mind is that you are pricing labor and material according to the time when the work is expected to be done, not according to the time the job is being estimated. Some electrical work is done several months after the bid is submitted. There's no way to be sure what prices will be in three to six months. Therefore, if you expect there will be a price escalation or labor shortage in the future, it is better to make some allowances now, or get some price guarantee in writing from your suppliers or subcontractors.

ESTIMATING MATERIAL COSTS

Basic Formula:
Material Price = Quantity × Material Unit Price

Do Your Own Take-off

You will always need to prepare your own material estimate. Since suppliers won't be doing the installation, their estimate may omit important items like supports and clips. Your own fixture/switchgear count is more likely to include everything that's needed to comply with the code and do the job.

Use a price book (the wholesale type) to plug in unit prices before getting material quotes. When the quotes come in later, you will have a baseline number to compare to. Figure 3.3 shows an example to work out a "plug" price.

Get Quotes

Always request quotes from at least three suppliers. Attach a copy of your take-off, even if they might use their own estimating services. Specify as much information as possible, including product type, model number and make, quantity, etc. Make a copy of electrical drawings and specs for them.

It is important not to play one supplier against the other. Win the trust of your supplier and try to ask him to give you his best price the first time.

ESTIMATING MATERIAL COSTS *(cont.)*

Evaluate Quotes

When quotes come in, read them carefully to verify the following:

- Unit Price
- Delivery Charge (you should look for FOB jobsite)
- Sales Tax
- Minimum Order Quantity
- Expected Price Escalation
- Storage Costs
- Discount Rate
- What is included
  (i.e. are lamps installed in lighting fixtures?)

Certain fixtures are not available to some suppliers and they may propose an alternate. It is hard to determine, at the time of the bid, whether an alternate is equal to the specified item. You can check with the design engineer to see if the alternate is acceptable, but some bargain items may require too much labor to install or too much effort to get approved.

Refer to Figure 3.4 for a worksheet you can use to evaluate all quotes in a matrix. If one supplier is out of line (i.e. too low), give him a call so that he might make corrections to all bidding electrical contractors. You should not mention any details regarding other quotes.

ESTIMATING LABOR COSTS

Basic Formula:
Total Man-hours = Quantity × Man-hour per Item
Total Labor Price = Total Man-hours × Labor
Hourly Rate

Man-hours

It is much harder to price labor than estimating material costs, because labor is subject to many variables. Typically labor is where most estimates go wrong.

Some electricians just apply a labor unit price rate (i.e. how much it costs to install each item) to their quantities, but unit prices are only as good as your experience using them especially when they change very often. What works on one job may not work out on the other. When you use those same money-making unit prices on another job, you can lose money. Though you feel that things were a little different from the last job, you don't know how to compensate for those differences.

A better way is to use man-hours (i.e. what a man can do within an hour). This information tends to remain relatively stable from job to job. For example, from previous jobs you have done, it seems to take you 3 hours to install a certain type of panel. You are currently paid $50.00 per hour. The labor to install two panels will cost: 2 EA × 3 man-hours × $50.00/man-hour = $ 300.

ESTIMATING LABOR COSTS *(cont.)*

A common mistake is to forget that man-hours vary for installation conditions. This book offers man-hours for some common electrical work items in Chapter 6. You should not use the information without adjusting them for specific job situations. Factors include:

- Job Size
- Overtime
- Size of Crew
- Delays/Interruptions
- Elevations
- Site Congestion
- Stacking of Trades
- Multiple Floors
- Weather
- Accessibility
- Pre-fabrication/Assembly

Chapter 6 offers more information on how to make adjustments to standard man-hours, or even calculate your own man-hours.

Labor Hourly Rate

In figuring out how much money you should charge per hour for your work (especially in those "time and material" jobs), it is important to remember the labor hourly rate should include:

- Basic wage
- Taxable fringe benefits (i.e. vacation pay)
- Worker's compensation insurance
- General Liability Insurance
- Medical insurance (health, dental, life, and disability)
- Living allowances and cash compensation
- Tax-deferred pension or profit sharing plans
- Social security and medicare taxes (FICA)
- Federal Unemployment Tax (FUTA)
- State Unemployment Tax (SUTA)
- Union Dues

Figure 3.5 gives an example to calculate labor hourly rate (excluding mark-ups) for a small company of four field crew members, based on the annual accounting records.

Figure 3.6 calculates the labor price, based on total man-hours needed to install that list of lighting fixtures and labor hourly rate.

ESTIMATING SUBCONTRACTOR COSTS

Quotes from subcontractors should be carefully read before deciding on a low number. You should have at least three quotes for each major specialty system.

Figure 3.8 gives an example worksheet to evaluate subcontractors' quotes. It is similar to compare material suppliers' quotes, but each material supplier priced from a close quantity take-off, while subcontractors may have different inclusion and exclusions. Thus it is important to pay attention to their scopes of work.

Sometimes telephone quotations from subcontractors or suppliers become unavoidable, especially at the last minute of the bid. Figure 3.9 offered a worksheet to record important details from telephone quotes. Fill in the form as completely as possible, and make as many notes as you need. List the job name, estimate number, the date and time the quote was received. Write the name of the company and of the person offering the quote. Find out if the price includes tax, freight and delivery to jobsite. Ask about anticipated price increases and exclusions. If you see any problems, make a note of them to be resolved later.

Overhead is one of your costs, not profit. You must pay or "recover" your overhead each year by doing enough business to cover its cost. You are not making any profits until all of the overhead costs are redeemed.

There are two types of overhead: jobsite overhead and office overhead. Each is calculated differently. Do not try to apply a flat rate (such as 10%) covering both, because sometimes that's not enough!.

Jobsite Overhead

Jobsite overhead refers to costs directly related to your specific job. It is true that general contractors pay for many such costs as toilet rental, material storage, etc., but inevitably you will have to pay certain jobsite overhead items out of your own pocket. Normally the longer the job, the more jobsite overhead costs will be.

Figure 3.10 provides a list of typical jobsite overhead cost items paid by electrical contractors. What you actually need to include depends on job conditions (or more). A large amount of jobsite overhead for electrical work is temporary power and lighting.

Jobsite overhead varies from job to job, and it is important to do an itemized estimate. (See Figure 3.11 for an example.)

ESTIMATING OVERHEAD COSTS *(cont.)*

Office Overhead

Office overhead may not be directly tied to a specific job. You must pay for these costs to remain in business, whether you have any jobs or not. Examples of office overhead items include:

- Owner's Salary
- Salaries and Fringe Benefits for Office Personnel (i.e. estimators, draftsmen, bookkeepers, secretaries, etc.)
- Non-Job Vehicles, Fuels, and Insurance
- Office Rent, Utilities, Postage, Furniture, and Supplies
- Small Tools and Equipment
- Business License and Professional Membership Dues
- Marketing and Advertising
- Loan Interests
- Legal Expenses
- Taxes and Donations
- Bad Accounts

Office overhead varies from year to year. The best bet is to look at how much work you did last year and the overhead you paid, figure a percentage based on that, and then apply that percentage to the current estimate. Refer to Figure 3.12 for an example of how to calculate office overhead costs.

ESTIMATING PROFIT

Profit is the money you want to make from the job and is normally estimated by applying a rate to total costs. The rate could run 20% to 30% for small jobs and 10% to 15% on a large one.

In deciding the rate to be used, it is important to have a clear understanding about what and who you are up against. Evaluate how much risk you are taking and how much money you could make. Then decide whether it's worth to take the gamble. If you think you want a job so badly and bid too low, chances are you will lose money.

One of the best ways to determine the profit you can expect in today's competitive market is to look at the trend on your completed jobs. Keep a chart of completed jobs handy (as shown in Figure 3.13), and evaluate each job's actual performance against the as-bid estimate. Use this information to decide what sort of profit you can hope for, and apply that rate to the current bid.

Profit is not a dirty word. Bid the project with a planned profit!

GETTING A CORRECT TOTAL

Now it is time to put together a correct total bid price. This is the "make-or-break" moment, as any cost items you missed will have to come out of your profit later.

Here is a quick run-down of the major cost areas:
- Material
- Labor
- Work by Subcontractors
- Jobsite Overhead (including equipment, etc.)
- Office Overhead
- Owner's Cash Allowance (check specs)
- Electrical Permit/License
- Bond and Insurance
- Contingency (bad design documents, material/labor escalation, unforeseeable field conditions)
- Profit

Shown in Figure 3.14 is a summary worksheet for figuring total job costs, both direct and indirect. Please keep in mind that the best way to avoid errors is to prepare an estimate as detailed as possible.

FIGURE 3.1 — PRICING PROCEDURES

1. Finish take-off first. Summarize the material quantities.

2. Combine the numbers for the same items and allow for reasonable waste.

3. Only transfer the total quantities to your pricing sheets.

4. Pricing materials based on the quotes you received from the suppliers.

5. Pricing labor based on the adjusted man-hour information.

6. Add indirect costs like overhead, bond, insurance, permits, etc.

7. Allow costs for items that are not shown on drawings but required.

8. Add owner's cash allowance for electrical work.

9. Allow contingencies due to problems in design and field construction.

10. Add profits to get a total price.

FIGURE 3.2 — PRICING MECHANICS

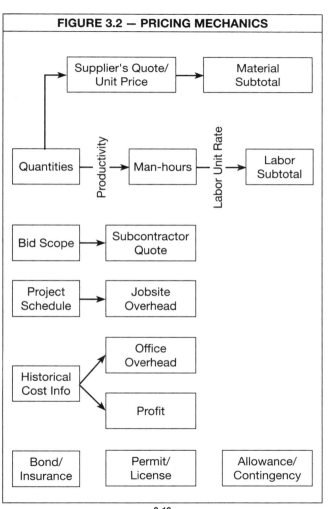

FIGURE 3.3 — MATERIAL PRICING WORKSHEET

Job Name: __ABC School__

Estimate No.: __901__   Estimator: __AD__

Date: __Jan. 1, 2008__   Worksheet Page Number: __P1__

Item	Quantity	Unit Price	Per	Extension
A	6	$ 130.00	E	$ 780.00
B	2	$ 40.00	E	$ 80.00
C	4	$ 50.00	E	$ 200.00
D	15	$ 135.00	E	$ 2,025.00
E	3	$ 50.00	E	$ 150.00
F	1	$ 90.00	E	$ 90.00
G	3	$ 100.00	E	$ 300.00
H	3	$ 25.00	E	$ 75.00
I	2	$ 35.00	E	$ 70.00
J	1	$ 40.00	E	$ 40.00
K	1	$ 69.00	E	$ 69.00
L	1	$ 75.00	E	$ 75.00
M	1	$ 180.00	E	$ 180.00
N	2	$ 90.00	E	$ 180.00
O	3	$ 45.00	E	$ 135.00
P	2	$ 40.00	E	$ 80.00
Q	6	$ 92.00	E	$ 552.00
			Subtotal	$ 5,081.00
		Sales Tax	6%	$ 304.86
		Freight	1%	$ 50.81
			Total Material Costs	$ 5,436.67

NOTE: "E" stands for "Each".

FIGURE 3.4 – SUPPLIER EVALUATION WORKSHEET

Fixture Type	Qty.	ACME Ltd. Unit Price	ACME Ltd. Extension	ABC Inc. Unit Price	ABC Inc. Extension	XYZ Corp. Unit Price	XYZ Corp. Extension
A	6	$ 125.00	$ 750.00	$ 130.00	$ 780.00	$ 130.00	$ 780.00
B	2	$ 30.25	$ 60.50	$ 35.25	$ 70.50	$ 36.25	$ 72.50
C	4	$ 40.50	$ 162.00	$ 45.50	$ 182.00	$ 46.50	$ 186.00
D	15	$ 121.00	$ 1,815.00	$ 126.00	$ 1,890.00	$ 127.00	$ 1,905.00
E	3	$ 40.00	$ 120.00	$ 45.00	$ 135.00	$ 46.00	$ 138.00
F	1	$ 79.00	$ 79.00	$ 84.00	$ 84.00	$ 82.00	$ 82.00
G	3	$ 94.85	$ 284.55	$ 92.85	$ 278.55	$ 90.85	$ 272.55
H	3	$ 19.95	$ 59.85	$ 17.95	$ 53.85	$ 15.95	$ 47.85
I	2	$ 31.00	$ 62.00	$ 29.00	$ 58.00	$ 27.00	$ 54.00
J	1	$ 34.25	$ 34.25	$ 32.25	$ 32.25	$ 32.75	$ 32.75
K	1	$ 64.00	$ 64.00	$ 62.00	$ 62.00	$ 62.50	$ 62.50
L	1	$ 71.00	$ 71.00	$ 69.00	$ 69.00	$ 69.50	$ 69.50
M	1	$ 175.00	$ 175.00	$ 173.00	$ 173.00	$ 173.50	$ 173.50
N	2	$ 83.00	$ 166.00	$ 84.00	$ 168.00	$ 82.25	$ 164.50
O	3	$ 35.00	$ 105.00	$ 36.00	$ 108.00	$ 35.00	$ 105.00
P	2	$ 38.00	$ 76.00	$ 39.00	$ 78.00	$ 38.00	$ 76.00
Q	6	$ 90.00	$ 540.00	$ 91.00	$ 546.00	$ 90.00	$ 540.00
Subtotal			**$4,624.15**		**$4,768.15**		**$4,761.65**
Sales Tax	6%		$ 277.45		$ 286.09		$ 285.70
Freight	1%		$ 46.24		Included		Included
TOTAL			**$4,947.84**		**$5,054.24**		**$5,047.35**

3-15

FIGURE 3.5 — ESTIMATING LABOR HOURLY RATE	
Field Employees	**Total Annual Wages**
John	$ 54,000
Adrian	$ 40,000
Tom	$ 35,000
Jack	$ 28,000
Subtotal	**$ 157,000**
Add:	
Bonus	$ 8,000
Living Allowances	$ 1,000
Social Security	$ 9,734
Medicare	$ 2,277
Federal Unemployment Tax (FUTA)	$ 224
State Unemployment Tax (SUTA)	$ 720
Worker's Compensation Insurance	$ 9,813
General Liability Insurance	$ 1,523
Health Insurance	$ 24,000
Dental Insurance	$ 4,800
Pension (401K)	$ 9,420
Union Dues	$ 1,440
Total Annual Payroll	**$ 229,951**

FIGURE 3.5 — ESTIMATING LABOR HOURLY RATE *(cont.)*				
Employee Name	**John**	**Adrian**	**Tom**	**Jack**
Hours Per Week	40	40	40	40
Weeks Worked	25	30	30	28
Total Working Hours	**1000**	**1200**	**1200**	**1120**
Vacations	3	2	1	0
Sick Leave	1	2	0	1
Total Non-Working Weeks	4	4	1	1
Hours Per Week	40	40	40	40
Total Non-Working Hours	**160**	**160**	**40**	**40**
Total Paid Hours	**1160**	**1360**	**1240**	**1160**

Thus the total hours are:
1,160 + 1,360 + 1,240 + 1,160 = 4,920 hours.

Because the annual payroll is $229,951, the average labor hourly rate is $229,951/$4,920 = $46.74, excluding any mark-ups. Round it up and use $50 per hour.

FIGURE 3.6 — LABOR PRICING WORKSHEET

Job Name: __ABC School__

Estimate No.: __901__   Estimator: __AD__

Date: __Jan. 1, 2008__   Worksheet Page Number: __P2__

Fixture Type	Quantity	Man-hour	Per	Extension
A	6	0.60	E	3.60
B	2	0.40	E	0.80
C	4	1.00	E	4.00
D	15	0.50	E	7.50
E	3	1.10	E	3.30
F	1	1.20	E	1.20
G	3	0.50	E	1.50
H	3	0.60	E	1.80
I	2	0.60	E	1.20
J	1	1.00	E	1.00
K	1	1.20	E	1.20
L	1	1.10	E	1.10
M	1	1.50	E	1.50
N	2	2.00	E	4.00
O	3	0.40	E	1.20
P	2	0.30	E	0.60
Q	6	0.25	E	1.50
		Total Man-hours		37.00
		Labor Hourly Rate		$50.00
		Burden		Included
		Labor Cost Subtotal		**$1,850.00**

NOTE: "E" stands for "Each".

FIGURE 3.7 — SALARY CONVERSION TABLE

Per Hour	Per Week	Per Month	Per Year
$6.00	$240	$1,039	$12,470
$7.00	$280	$1,212	$14,549
$8.00	$320	$1,386	$16,627
$9.00	$360	$1,559	$18,706
$10.00	$400	$1,732	$20,784
$11.00	$440	$1,905	$22,862
$12.00	$480	$2,078	$24,941
$13.00	$520	$2,252	$27,019
$14.00	$560	$2,425	$29,098
$15.00	$600	$2,598	$31,176
$16.00	$640	$2,771	$33,254
$17.00	$680	$2,944	$35,333
$18.00	$720	$3,118	$37,411
$19.00	$760	$3,291	$39,490
$20.00	$800	$3,464	$41,568
$21.00	$840	$3,637	$43,646
$22.00	$880	$3,810	$45,725
$23.00	$920	$3,984	$47,803
$24.00	$960	$4,157	$49,882
$25.00	$1,000	$4,330	$51,960
$26.00	$1,040	$4,503	$54,038
$27.00	$1,080	$4,676	$56,117
$28.00	$1,120	$4,850	$58,195
$29.00	$1,160	$5,023	$60,274
$30.00	$1,200	$5,196	$62,352
$31.00	$1,240	$5,369	$64,430
$32.00	$1,280	$5,542	$66,509
$33.00	$1,320	$5,716	$68,587
$34.00	$1,360	$5,889	$70,666
$35.00	$1,400	$6,062	$72,744
$36.00	$1,440	$6,235	$74,822
$37.00	$1,480	$6,408	$76,901
$38.00	$1,520	$6,582	$78,979
$39.00	$1,560	$6,755	$81,058
$40.00	$1,600	$6,928	$83,136
$41.00	$1,640	$7,101	$85,214
$42.00	$1,680	$7,274	$87,293
$43.00	$1,720	$7,448	$89,371
$44.00	$1,760	$7,621	$91,450
$45.00	$1,800	$7,794	$93,528
$46.00	$1,840	$7,967	$95,606
$47.00	$1,880	$8,140	$97,685
$48.00	$1,920	$8,314	$99,763
$49.00	$1,960	$8,487	$101,842
$50.00	$2,000	$8,660	$103,920
$51.00	$2,040	$8,833	$105,998
$52.00	$2,080	$9,006	$108,077
$53.00	$2,120	$9,180	$110,155

NOTE: This table is based on a 40-hr. week, 4.33-week month of a 52-week year.

FIGURE 3.7 — SALARY CONVERSION TABLE (cont.)

Per Hour	Per Week	Per Month	Per Year
$54.00	$2,160	$9,353	$112,234
$55.00	$2,200	$9,526	$114,312
$56.00	$2,240	$9,699	$116,390
$57.00	$2,280	$9,872	$118,469
$58.00	$2,320	$10,046	$120,547
$59.00	$2,360	$10,219	$122,626
$60.00	$2,400	$10,392	$124,704
$61.00	$2,440	$10,565	$126,782
$62.00	$2,480	$10,738	$128,861
$63.00	$2,520	$10,912	$130,939
$64.00	$2,560	$11,085	$133,018
$65.00	$2,600	$11,258	$135,096
$66.00	$2,640	$11,431	$137,174
$67.00	$2,680	$11,604	$139,253
$68.00	$2,720	$11,778	$141,331
$69.00	$2,760	$11,951	$143,410
$70.00	$2,800	$12,124	$145,488
$71.00	$2,840	$12,297	$147,566
$72.00	$2,880	$12,470	$149,645
$73.00	$2,920	$12,644	$151,723
$74.00	$2,960	$12,817	$153,802
$75.00	$3,000	$12,990	$155,880
$76.00	$3,040	$13,163	$157,958
$77.00	$3,080	$13,336	$160,037
$78.00	$3,120	$13,510	$162,115
$79.00	$3,160	$13,683	$164,194
$80.00	$3,200	$13,856	$166,272
$81.00	$3,240	$14,029	$168,350
$82.00	$3,280	$14,202	$170,429
$83.00	$3,320	$14,376	$172,507
$84.00	$3,360	$14,549	$174,586
$85.00	$3,400	$14,722	$176,664
$86.00	$3,440	$14,895	$178,742
$87.00	$3,480	$15,068	$180,821
$88.00	$3,520	$15,242	$182,899
$89.00	$3,560	$15,415	$184,978
$90.00	$3,600	$15,588	$187,056
$91.00	$3,640	$15,761	$189,134
$92.00	$3,680	$15,934	$191,213
$93.00	$3,720	$16,108	$193,291
$94.00	$3,760	$16,281	$195,370
$95.00	$3,800	$16,454	$197,448
$96.00	$3,840	$16,627	$199,526
$97.00	$3,880	$16,800	$201,605
$98.00	$3,920	$16,974	$203,683
$99.00	$3,960	$17,147	$205,762
$100.00	$4,000	$17,320	$207,840

NOTE: This table is based on a 40-hr. week, 4.33-week month of a 52-week year.

FIGURE 3.8 — SUBCONTRACTOR EVALUATION WORKSHEET

System Quoted	Plug Qty./ Price	Sub #1	Sub #2	Sub #3
Communication System				
Base Price	—	$15,000	$14,000	$16,000
Work Included	Telephone	Yes	Yes	Yes
	Intercom	No	No	No
	Cable	Yes	Yes	Yes
	Internet	Add $2,000	Add $2,000	Incl.@ 2,000
Delivery Time	—	N/A	N/A	N/A
Adjustments	—	—	—	—
Sales Tax	—	Included	Add $840	Included
Delivery to Jobsite	—	Included	Included	Included
Complete Installation	—	Included	Included	Included
Per Plans and Specs	—	Yes	Yes	Yes
Exclusions	—	None	None	None
Addenda Received	—	Yes	Yes	Yes
Other Factors	—	N/A	N/A	N/A
Adjusted Total Bid	—	$17,000	$16,840	$16,000

FIGURE 3.9 — TELEPHONE QUOTATION WORKSHEET

Date:_____

Project Name:_____

Quote Taken By:_____

Subcontractor/Supplier Name: _____

Contact Person: _____

Telephone:_____

Fax:_____

Addendum Received: _____

Scope of Work:_____

Proposal Amount: _____

Alternates: _____

Inclusions: _____

Exclusions: _____

Sales Tax: _____Yes _____No

Delivery to Job Site: _____Yes _____No

Installation: _____Yes _____No

Per Plans and Specs: _____Yes _____No

Adjusted Total Amount: _____

FIGURE 3.10 — ELECTRICAL JOBSITE OVERHEAD ITEMS

__ Job mobilization and demobilization

__ Tool sheds

__ Material storage trailer

__ Blueprints for subcontractor and supplier

__ Supervision

__ Travel expenses

__ Job vehicles and fuels

__ Temporary power and lighting

__ Scaffolding or lifts for electrical work

__ Jobsite signs

__ Trade clean-up

__ Shop drawings

__ Product sample submittals

__ Punch-list Items

__ As-builts

__ Callback during warranty period

FIGURE 3.11 — ESTIMATING JOBSITE OVERHEAD

This is the sample jobsite overhead estimate for a medium-size electrical job:

Item	QTY	Unit	Rate	Subtotal
Mobilization	1	l/s	$400	$400
Superintendent	8	wk	$1,500	$12,000
Foreman	8	wk	$800	$6,400
Hotel expenses	8	wk	$500	$4,000
Project sign	1	l/s	$500	$500
Telephone bills	2	mo	$150	$300
Temporary power	1	l/s	$5,000	$5,000
Vehicles and fuels	2	each	$2,000	$4,000
Hoisting	1	wk	$3,000	$3,000
Small tools	1	l/s	$500	$500
Submittals	1	l/s	$400	$400
Daily clean-up	40	days	$80	$3,200
Punch-list items	1	l/s	$1,000	$1,000
As-built drawings	1	l/s	$300	$300
Total Jobsite Overhead				**$41,000**

FIGURE 3.12 — ESTIMATING OFFICE OVERHEAD

Formula:

Rate = Office Overhead last year/Construction Volume last year

Total Direct Costs for current bid = Material + Labor + Subcontractor + Jobsite Overhead

Office Overhead for current bid = Rate × Total Direct Costs for current bid

Example:

Office Overhead last year: $30,000

Construction Volume last year: $500,000

Office Overhead rate: $30,000/$500,000 = 6%.

Total Direct Costs for your current bid: $190,000

Office Overhead for current bid:
$190,000 × 6% = $11,400

FIGURE 3.13 — ESTIMATING PROFIT RATE

Job Name	Contract Amount	Job Superintendent	As-Bid Profit Rate	Actual Profit Rate	Rate Evaluation
Today's Rate					

FIGURE 3.14 — PRICING SUMMARY WORKSHEET			
Item	**Rate**	**Amount**	**Subtotal**
Material Quotes			
Switchgear			$3,100
Fixtures and Lamps			$5,048
Conduit and fittings			$1,500
Boxes and Wires			$1,000
Devices			$800
Hook-up			$500
Miscellaneous			$1,200
Sales Tax	6%		$789
Material Subtotal			**$13,937**
Labor Subtotal			**$20,000**
Subcontractor Subtotal			**$15,000**
Jobsite Overhead			**$10,000**
Total Direct Costs			**$58,937**
Office Overhead	5%		$2,947
Owner's Allowance			$1,500
Bond/Insurance			$2,100
Electrical Permit			$1,200
Contingency			$3,000
Profit	15%		$10,228
Total Bid Price			**$78,412**

FIGURE 3.15 – SHORTCUTS FOR ESTIMATING MARK-UP

	Office Overhead Rate										
Profit Rate	15%	16%	17%	18%	19%	20%	21%	22%	23%	24%	25%
5%	0.80	0.79	0.78	0.77	0.76	0.75	0.74	0.73	0.72	0.71	0.70
6%	0.79	0.78	0.77	0.76	0.75	0.74	0.73	0.72	0.71	0.70	0.69
7%	0.78	0.77	0.76	0.75	0.74	0.73	0.72	0.71	0.70	0.69	0.68
8%	0.77	0.76	0.75	0.74	0.73	0.72	0.71	0.70	0.69	0.68	0.67
9%	0.76	0.75	0.74	0.73	0.72	0.71	0.70	0.69	0.68	0.67	0.66
10%	0.75	0.74	0.73	0.72	0.71	0.70	0.69	0.68	0.67	0.66	0.65
11%	0.74	0.73	0.72	0.71	0.70	0.69	0.68	0.67	0.66	0.65	0.64
12%	0.73	0.72	0.71	0.70	0.69	0.68	0.67	0.66	0.65	0.64	0.63
13%	0.72	0.71	0.70	0.69	0.68	0.67	0.66	0.65	0.64	0.63	0.62
14%	0.71	0.70	0.69	0.68	0.67	0.66	0.65	0.64	0.63	0.62	0.61
15%	0.70	0.69	0.68	0.67	0.66	0.65	0.64	0.63	0.62	0.61	0.60
20%	0.65	0.64	0.63	0.62	0.61	0.60	0.59	0.58	0.57	0.56	0.55
25%	0.60	0.59	0.58	0.57	0.56	0.55	0.54	0.53	0.52	0.51	0.50
30%	0.55	0.54	0.53	0.52	0.51	0.50	0.49	0.48	0.47	0.46	0.45
35%	0.50	0.49	0.48	0.47	0.46	0.45	0.44	0.43	0.42	0.41	0.40
40%	0.45	0.44	0.43	0.42	0.41	0.40	0.39	0.38	0.37	0.36	0.35
45%	0.40	0.39	0.38	0.37	0.36	0.35	0.34	0.33	0.32	0.31	0.30
50%	0.35	0.34	0.33	0.32	0.31	0.30	0.29	0.28	0.27	0.26	0.25

Estimating Example:
If the direct costs (labor, material, and job overhead) add up to $25,000

• You want to apply 18% of total bid as office overhead
• You want to apply 10% of total bid as profit
• From the table, the "mark-up" rate is 0.72
• The result can be found by: $25,000/0.72= $34, 722
• You still need to add owner's cash allowance, bond, insurance, permit, and contingency to get the final bid price.

CHAPTER 4
Bidding

Bid day can be quite hectic, both for those who prepare the price proposals and those who receive them. No matter how prepared you think you are, there could be surprises, if not problems. Being more organized could reduce errors, as well as stresses.

WRITING A PROPOSAL

After figuring out a total bid price, you are now ready to submit a proposal to the general contractor or owner. Choose the words of your proposal carefully. An example proposal form is provided in Figure 4.1.

Include the following information on your proposal:
- Proposal number, date, and version of revision
- Complete job name and address
- Name of owner, architect/engineer, or general contractor
- A complete list of electrical drawings and their issue or revision date
- Electrical specification titles and issue date
- A complete list of addenda and their dates of issue
- Base bid price as well as the numbers for alternates

- A list of inclusions, exclusions, clarifications, and assumptions
- A standard contract from your company
- Your contact name and phone number
- How long the price is good for

You are expected to bid your section as it is specified. General contractors will ask you if your bid covers certain hookups or wiring listed in other sections of the job, i.e. power for irrigation controllers, etc. Here's what to do:

1. If you feel certain items are questionable and would not normally be covered under your section, be prepared to exclude them from your proposal.
2. If you find something is questionable and might be covered under your section, be prepared to include that item in your bid.

Figure 4.2 through 4.7 gives a standard list of inclusions and exclusions for electrical contractors. Determine what areas of your work are best handled by the general contractor. Unless otherwise specified, excluding items such as trenching, placing concrete pads, and painting can lower your overall price. It is always better to double check plans and specs to see what is supposed to be done by electricians on your current job.

COST BREAKOUT REQUIREMENTS

Owners normally put their cost breakdown requirements into the specs. For example, if there are two buildings, they want to know the cost for each. Electrical contractors bidding on a job should verify such requirements with general contractors or owners, and prepare proposals accordingly.

Sometimes you either forget, or are not informed about the need for cost breakout until you are done with all estimating and pricing. If on the bid day the customer wants your total prices to be broken out into several smaller pieces, there are still a couple of ways you can break up the numbers.

A common basis is breaking out by the gross building area (i.e. square feet). Suppose there are quite a few buildings in one job. It normally makes sense that the larger buildings should have a larger share of the total costs, if the other buildings are quite similar.

Another way is to break out by the number of functional components such as the residential condo units, hotel rooms, hospital beds, etc.

Example calculations are shown in Figure 4.8.

ASSESSING RISKS

You have completed your take-off, entered the quotes, applied overhead and profit, and written a clear proposal with inclusions and exclusions. You believe you have accounted for everything. What could possibly go wrong? Plenty!

The question is this: Where will the project go wrong and what can you do as the estimator to have its risks covered as much as possible?

Have a review of the following factors:

- Duration of the project
- Number of labor and total hours required
- Potential hazards
- Weather conditions
- Potential delays caused by failure of other trades
- Possible procurement failures (i.e. owner furnished lighting package)
- Amount of engineering and drafting required
- Quantity of construction equipment and tools required
- Materials procurement cost
- Labor costs and its percentage in relation to the total price

Now go back to the fundamental question: Should you take on this job? You thought about it before deciding to bid on it and now it's time to refresh the answer, especially now that you are more familiar with the job.

- Will the job disrupt or interfere with other operations of your company?
- Is your company trained and equipped for this kind of job?
- Are there sufficiently trained electricians readily available?
- Is the present stock of tools and equipment adequate for this type of job?
- Are there other electrical contractors available who are more able to handle the work?

Discuss with your field superintendents or foremen who did similar projects previously. Identify any problems they may have encountered. Review the job schedule with them and listen to what they have to say.

Eventually, you need to see clearly where the risks are and how much they might cost you individually if things go wrong. By adding all these "risk-dollars", you can decide if it is worth to gamble. Figure 4.9 shows you an example of how to do that.

Figure 4.10 provides a checklist to follow, before you send out a proposal. Your final bid can be phoned in, sent by fax, or by e-mail if it is in an electronic format. Sending by regular mail is normally too slow and the proposal can get lost.

If you need to offer a bid by telephone, Figure 4.11 gives some advice. Always follow up with a written quotation listing what you discussed in the phone conversation.

Before the bid day, it is a good idea to send out a "scope quote" with everything completed except for the price. This is to assure the general contractor that you are still bidding and will send a formal proposal the next day. It also gives him some clues about what you will be bidding on. For example, if you don't want to include temporary power and lighting, then he will know that in advance and make allowances to cover it.

Try to turn in your proposal before the deadline required. If you cannot make a bid or need more time, please call the general contractor or the owner as soon as possible to discuss it. Do not wait until the last minute.

It is a good idea to call the receiving party after your proposal is sent to make sure they have received it. Get the name of the estimator who will evaluate the electrical quotes. Make yourself available on the bid day to answer questions.

WHAT TO DO WITH BID SHOPPING

It is true that in today's construction industry, there are some owners and general contractors who do not run their business morally. They can use your already low quote to solicit lower bids from your competition, or pressure you into cutting your price by threatening to award the work to others.

This unethical practice is called "bid shopping", and actually can occur either during or after the bid (when the general contractor has the job). When the bid process does not require the general contractor to list the subs he plans to use, bid shopping is more common.

Some electrical contractors become part of "bid shopping", partly because they want the job too much. They inquire about details regarding their competition's quote, and then offer to beat that price. Others simply guarantee to provide a figure which is a certain percentage below the current, lowest bid, however, a job that is profitable for one contractor may be unprofitable for the other. Essentially, you are making a gamble in cutting the price.

The real solution is to know whom you are bidding with. It's a waste of time to bid for general contractors who shop their sub bids. On bid days, some general contractors may suggest that your price was too low or too high compared to other bids, but bid days don't allow much time for checking every worksheet. If you are not sure of your price, withdraw the bid.

FIGURE 4.1 — BID PROPOSAL FORM

Proposal No.: _____ Revision No. _____

Date: _____

From: _____

To: _____

Project Name and Address: _____

_____

We hereby propose to furnish materials and perform labor necessary for the construction of electrical systems for the above project. The proposal is based on electrical drawings _____and specifications _____ prepared by _____ and dated _____.

Receipt of Addendum # _____ is acknowledged.

Base Bid: _____ Dollars ($_____)

Alternate: _____ Dollars ($_____)

Our bid includes: _____

Our bid excludes:_____

This proposal is valid for _____days from the date of the submission.

Respectfully submitted by: _____

Signature: _____

Title: _____

FIGURE 4.2 — GENERAL PROPOSAL SCOPE INCLUSIONS

- Contingency funds specifically called for in the specifications
- Cash allowances defined in the specifications
- All service charges as levied by the town, city, county, or utility company
- Sales tax
- Electrical permits and licenses
- General liability insurances
- Trade clean-up
- Protection of other trades' work from damage
- Rigging, hoisting, and placing of equipment or materials
- Scaffolding required by the work of electrical trade
- Demolition or relocation of existing electrical system components
- Cut penetrations of 6" and smaller through walls, floors, and ceilings
- Fire-stopping materials at drilled holes, sleeves, and other openings
- Patching and making good materials related to openings under this trade
- Shop drawings, product samples, and submittal data
- As-built drawings, maintenance data, and operating instructions to owners
- Fastening devices, supports, and frames for electrical equipment.
- Identification, adhesive markers, tags, labels, and framed directories.
- Access panels for servicing of electrical equipment
- Electrical load balancing

FIGURE 4.3 — DISTRIBUTION PROPOSAL SCOPE INCLUSIONS

- Guarantees and warranties
- Permanent electrical service to utility connection point substations
- Transformer vaults including related wiring and equipment
- Main and secondary switchboards
- Bus duct
- Branch circuit feeders
- Meters
- Generator sets
- Emergency power panels
- Motor control centers
- UPS (Uninterruptible Power Supplies)
- Grounding systems
- Lightning protection systems
- Supply and installation of electric sub-distribution (breakers, splitter bars, enclosures) for air handling units

FIGURE 4.4 — WIRING PROPOSAL SCOPE INCLUSIONS

- Conduit, wire flexible
- Empty conduit systems
- Cable trays and raceways
- Floor duct systems (including cutting of holes)
- Motor starters and protection
- Contactors
- Lighting protection systems
- Interrupter switches and load shedding devices
- Dimmer switches
- Wiring devices, switches, receptacles, and plates
- Multi-speed switches
- Rectifiers, regulators, relays, and resistors
- Ground fault devices
- "Trade name" manufactured car heater posts and lamp standards
- Supply and installation of contactors required for electric duct heaters
- Installation and power wiring of mechanical equipment controls, if not done by others
- Wiring and connection for electrical appliances, mechanical package units, and illuminated commercial building signage

FIGURE 4.5 — FIXTURES PROPOSAL SCOPE INCLUSIONS

- Lighting fixtures and lamps
- Luminous elements consisting of a light source (ballasts if applicable) mounted separately or as components of an overall ceiling system
- Fixtures, lamps, and lenses for coffered ceilings
- Lighting fixtures and lamps for luminous ceilings and light valances
- Air troffers and indoor or outdoor flood lighting
- Lenses, globes, guards, diffusers, and supports that are part of the lighting fixtures
- Directional illuminated signs
- Emergency lighting and battery equipment
- Installation of electric hand dryers, operating room lights and X-ray viewers, etc.

FIGURE 4.6 — SPECIAL SYSTEMS PROPOSAL SCOPE INCLUSIONS

- Heating equipment:
  - Baseboard heaters
  - Heat cables, tapes, and snow melting cables
  - Unit heaters and re-heat coils (electric)
  - Force flow heaters (electric) independent of ductwork connection
  - Self-contained electric heating devices
  - Terminal temperature control devices
- Detector systems: burglar, intrusion alarm
- Clock systems: time recording, watchmen
- Access systems: security, surveillance
- Fire alarm systems: heat, smoke detections, sprinkler annunciation
- Sound Communications Systems
  - Intercommunications systems: private telephone, wireless
  - Nurse call systems: doctor register
  - Public address systems: music, paging
  - Television systems: distribution, monitoring
  - Electronic learning systems: audio/visual

FIGURE 4.7 — PROPOSAL SCOPE EXCLUSIONS

- Any and all portion of the building permit
- Major material price increases
- Engineered design of electrical systems in addition to bid documents
- Errors in design documents
- Performance or payment bonds
- Delays or defaults due to strikes, accidents, fires, acts of god, or any other unavoidable causes beyond the contractor's reasonable control
- Temporary heat, water, sanitation
- Temporary light and power, unless specified under electrical trade
- Project trailer or office facilities
- Project insurance
- Jobsite security and hoarding
- Project safety and first-aid
- General survey and layout
- Reasonable access to site, including snow clearing
- Distribution and review of samples, shop drawings, and other submissions
- Protection of building finishes (i.e. floors, wall surfaces, ceilings, etc.).
- Independent inspections and testing
- Material storage costs
- Hauling of site trash
- Final clean-up
- Supports, backings and guards not shown on electrical drawings

FIGURE 4.7 — PROPOSAL SCOPE EXCLUSIONS *(cont.)*

- Asbestos removal
- Work related to underground obstructions or utilities beyond the scope of bid documents
- Over excavation of soft spots and backfill of the same
- Wiring for electric elevators, hydraulic elevators, parking garage hoists, escalators, dumbwaiters, moving walkways, and speed ramps
- Luminous ceilings, light valances, and coves
- Control wiring for equipment under mechanical scope
- All cast-in-place, concrete rebar, and forming including required excavating
- Painting, priming, and surface preparation
- Plaster, caulking, and grouting of all types
- Structural cutting, penetration, patching, or repairing
- Catwalks, ladders, and grating, unless required to be done by this trade
- Fireproofing or fire-stopping, unless required to be done by this trade
- Supply or installation of FFE (furniture, fixture, and equipment) unless required to be done by this trade
- All residential appliances
- Food service equipment, unless specified under electrical trade
- All modular unit components
- Car posts when part of railing, wheel stops, and bumper rails
- Commissioning

FIGURE 4.8 — COST BREAKOUT CALCULATION

1. Based on Building Area

Estimating Example:

You total bid price is $50,000.

Building A has 1,500 SF and Building B has 2,500 SF

Then the total building area is:

1,500 + 2,500 = 4,000 SF

Percentage for Building A of the total area:

1,500/4,000 × 100% = 37.5%

Percentage for Building B of the total area:

2,500/4,000 × 100% = 62.5%

Therefore

Cost for Building A: 37.5% × $50,000 = $18,750

Cost for Building B: 62.5% × $50,000 = $31,250

2. Based on Functional Components

Estimating Example:

You total bid price is $600,000.

Building A has 40 condo units each, and Building B has 60 condo units

Then the total condo units are: 40 + 60 = 100 tons

Percentage for Building A of the total condo units:

40/100 × 100% = 40%

Percentage for Building B of the total condo units:

60/100 × 100% = 60%

Therefore

Cost for Building A: 40% × $600,000 = $240,000

Cost for Building B: 60% × $600,000 = $360,000

FIGURE 4.9 — ESTIMATING "RISK-DOLLARS"

Estimating Example:

You found out the fire alarm and CCTV systems are the two major areas where you lost money in recent institutional projects.

You are now bidding a small school job about $120,000 in total electrical price.

The cost for the fire alarm system is about $7,000 and CCTV $4,000. Your typical loss is about 50% and 40% respectively.

Then your "Risk Dollars" are

Fire alarm system: $50\% \times \$7,000 = \$3,500$

CCTV system: $40\% \times \$4,000 = \$1,600$

Total "Risk Dollars" estimated
$\$3,500 + \$1,600 = \$5,100$

Your revised total base bid is
$\$120,000 + \$5,100 = \$125,100$

The percentage of "Risk Dollars" in your total bid is
$\$5,100/\$125,100 \times 100\% = 4.08\%$

FIGURE 4.10 — QUESTION LIST
BEFORE SENDING A BID

1. Have you double-checked the time and date for the bid to be submitted?
2. Does your bid include everything required in the scope?
3. Is your bid in accordance with plans and specs?
4. Is your bid in accordance with electrical codes?
5. Does your bid exclude everything that should not be included?
6. Are you aware of project schedule requirements? Can you finish it on time?
7. If you worked the bid from an old estimate, have you made applicable changes?
8. Did you check your math from beginning to end?
9. Have your questions been answered by the architect, engineer, or general contractor?
10. Have you received and reviewed all addenda transmitted by general contractor?
11. Have you checked with electrical engineers to see if you have all electrical-related addenda?
12. Did you breakout your quote in the format as required by the bid?
13. Are there any gaps in the quotes from your subs that you need to fill on your own?
14. Did you include bond, insurance, and electrical permit?
15. Did you include required owner's cash allowance?
16. Did you include some contingencies for design issues, price escalation, and unforeseeable site conditions?
17. Did you fill out all the blanks including alternates as well as base price?
18. Is the bid proposal signed by the company owner?
19. Do you want to add a business letter or standard contract?
20. Can the general contractor reach you on time if they have any questions regarding the bid?
21. Did someone else double check your estimate and proposal?

FIGURE 4.11 — CHECKLIST FOR TELEPHONE BIDDING

Speak slowly and clearly. Tell the general contractor the following information:

- Name of the project
- Name of your company
- The bid price in full dollar amount
- Your name
- Your telephone number
- The spec sections you are bidding
- Whether or not the bid is in accordance with plans and specs
- Whether or not the bid includes taxes, fees, and installation
- Any additional items included
- All exclusions from the spec sections you are quoting
- Addenda received
- Any alternate bid price as required

Write down the following information:

The name of the general contractor you bid to:

_____

The price you give to him: _____

The name of the person you speak with:

_____

You need to ask the other party to repeat the entire bid over the phone and correct any errors he may have made.

FIGURE 4.12 — QUESTION LIST FOR RENOVATION JOBS

- Are old drawings or specs (including as-built) for the existing building available?
- Which general contractor and electrician built the existing structure?
- Where is existing power line located? Is it difficult to gain access?
- Is the building poorly or well maintained?
- Are there any unusual job conditions (i.e. high ceilings, flooded basement, crawl space, occupant use, etc.)?
- Does the job involve 12 foot section cable trays in an existing drop ceiling? Who will remove and replace the ceiling grids?
- Does the job require working over and around machinery, office furniture, and drop ceilings or in confined areas?
- Is it difficult to move tools, equipment, and material around?
- Are exiting fixtures, conduits, and equipment to be removed or relocated?
- Are cutting and patching of existing surfaces part of electrical contract?
- Any pavement cutting and repairing required?
- Are there any dust control and noise abatement requirements?
- Are there any working-hour restrictions?
- Is there a problem with storage space on site?
- Have you visited the site to verify current condition and locate existing utilities?
- Have you increased labor and material rates to cover the unknowns?

CHAPTER 5
Post-Bid

Estimating does not end with faxing a copy of a proposal to the customer. A lot of things can happen after you sent a bid. Do not sit and wait for the phone to ring. Instead, be proactive and you may land a great job!

POST-BID REVIEW

Immediately after the bid is over, sit down to have a review of the bid you submitted. Check for any areas that you could work on improving before your next bid. Figure 5.1 and 5.2 listed some common estimating mistakes and ways to reduce them.

For each job you bid, some analytical work with the price proposal will prove to be very useful in the long run. For example, after bidding the same type of projects several times, you will have a better idea about the cost per square foot, etc. A worksheet is provided in Figure 5.3 for this purpose.

It is also important to do a little record-keeping as you never know when the customer is going to call back with questions. Organize all the paperwork in a post-bid folder for future reference as listed in Figure 5.4.

PROPOSAL FOLLOW-UP

Call the general contractor approximately two weeks after the bid to do a follow-up. Figure 5.5 listed some questions you can ask.

E-mails can also be used as an alternative to phone calls in inquiring the status. Some people are more willing to respond to e-mails rather than phone calls, as e-mails do not interfere what they are doing.

It might be difficult to arrange for a face-to-face meeting, which is better than a phone conversation. Do whatever it takes, including just showing up at their office and waiting in the lobby until they have time to meet you. Be well prepared to make a good presentation.

The project you bid might be over budget, and you could be requested to participate in a "value engineering" exercise (i.e. the cost saving by changing the design). For example, if you believe a smaller panel than the one currently designed could be equally used to meet the demand of building, it could result in some cost savings.

Figure 5.6 offers some common value engineering ideas. It is important that the revised design you are proposing is compliant with governing electrical code so that the job will pass final inspection.

IMPROVE BID-HIT RATIO

Studies show that electricians often bid six or seven jobs before they get one. If you get every job you bid on, you are bidding too low. On the other hand, if only one bid turns out to be successful for every 20 or 30 jobs you bid, you are spending too much time and effort estimating, and not enough time working.

Most electricians just keep bidding to the same customers over and over, using the same strategy as they've always used. At some point, they may realize this is not the best way to do it.

A little research can go a long way to make your bid more profitable. Develop a personal bid history and track it monthly, quarterly, and yearly. Track all types of projects you bid on and each customer you bid to: large versus small, hard bid versus negotiated, plans-specs versus design-build, new construction versus renovation, local versus out of town, commercial versus residential versus industrial, public versus private, etc.

As you study these jobs, you'll find certain customers give you more work than others. You'll also discover you do better with certain kinds of jobs. This simple tracking method will help you to focus on the real customers who can give you jobs. For some jobs, you will be surprised to find out price is not the most important factor. Figure 5.7 offers a preliminary worksheet for you to use.

ESTIMATING FOR CONTRACT

If the general contractor is talking to you regarding your proposal, you know at least your price is competitive, and maybe you are the low bidder. But before offering or accepting a subcontract, a revised estimate needs to be done. Rarely is the final electrical contract 100% based on initial bid proposal. There are almost always numerous post-bid changes to consider. Use the guidelines in Figure 5.9 for preparing a revised post-bid estimate.

Finally you are ready to sign a contract. Be aware about what you are signing. If you have your own "standard" contract, see if the general contractor or the owner is willing to accept it as an alternative to their contract.

If this is not possible, thoroughly read the contract they insist on. Read everything including the fine print. Do not assume that this contract is the same as the one in the original bid package. Verify it word for word. If something seems unfair to you, offer to cross it out. If they claim that questionable portion is not relevant, you should reply that it is better then to eliminate it completely. If you want something that isn't there, write it on all copies and have them initial the addition.

Figure 5.10 gives major elements in an electrical subcontract, while in Figure 5.11 there is an example work scope.

In addition, you need to pay special attention to the following contract items:

- Payment Clauses

- Penalties and Indemnity Clauses

- Notice Provisions

- Change Order Procedures

- Dispute Resolution

- Breach of Contract

- Cancellation of Contract

- Mechanics Lien

- Payment and Performance Bonds

- Time Limit of Legal Rights

Figure 5.12 lists some negative clauses you should always avoid. In any event, do not start the work until you have your copy of the amended, countersigned, and initialed contract in your possession.

TURN-OVER MEETING

If you are lucky enough to get a seemingly profitable job, then the next thing to do is hold a "turn-over" meeting with your field personnel who will run the daily jobsite operation.

Why have "turn-over" meetings when everyone can just look at a copy of the job estimate for answers? Because project managers and superintendents are much better informed about the project by communicating with the estimator, who knows the job best. Then they are able to start the submittal process and place vendor and subcontractor purchase orders accurately, and provide all the necessary materials and tools for the quick starts.

To have a successful meeting, determine who should attend: estimator, project manager, coordinators, superintendents, jobsite trade foreman, and job accountant. External personnel are normally not permitted, as "turn-over" meetings always involve sensitive information such as contract amount, low suppliers and subcontractors, etc.

A "turn-over" meeting should be more than just handing over a package of plans and specs. Instead, follow a meeting agenda like the one in Figure 5.13.

ESTIMATING FOR PROJECT MANAGEMENT

After construction started for the project, the estimator can still help project managers with change order pricing, job costing, etc.

It's important to have a list of cost codes to control and monitor actual construction expenses. You need to break down your estimate into smaller components and assign them different cost codes or account numbers. A simple list of no more than 10 cost codes will normally be enough. Then you can have the field foreman report labor and material expenditures by the cost codes. You can use industry standard cost codes or simplified ones as shown in Figure 5.14.

Some advice to handle change orders is provided in Figure 5.15. In today's construction world, if the job goes badly, you need to have something to protect yourself. Record-keeping is essential, including phone conversations. Figure 5.16 gives you a worksheet for that purpose.

Estimating for project management provides estimators with the ability to track actual costs versus estimated costs. Knowing the collected job cost data and how it compares to the estimated costs are worth more to an estimator than all the data gathered from other sources. Such knowledge takes the mystery out of labor units or material buying levels, and enables the estimator to make adjustments for future cost estimates.

UNIT PRICING

Ideally, your take-off will list every conduit, wire, strap, plate and box, etc., that goes into the job. But sometimes you might run out of time and it becomes impossible to measure everything.

By keeping complete cost information on every job they do, experienced electricians often have a good idea of how much it takes to install one linear foot of conduit, including wire, couplings, connectors, straps, etc. Related materials are grouped into assemblies that can be priced as a unit, without sacrificing accuracy. This is called unit pricing. Figure 5.17 gives you an example for unit price estimating.

The key for this method is to work out a correct unit price for each item required for pricing. Maintaining a historical job cost database is essential to develop and update unit prices periodically. Figure 5.18 provides a completed worksheet to work out the unit price for one commercial type duplex receptacle, material, installation, overhead, and profit included. Then all you have to do is to count the duplex receptacle on drawings and multiply the unit price you calculated.

Unit pricing is very useful to estimate "design-build" jobs, where very little information is shown in plans and specs (if they exist) on electrical work. Figure 5.19 provided some suggestions for estimating "design-build" projects.

FIGURE 5.1 — COMMON BIDDING MISTAKES

- **Not including required items.** A scope omission is perhaps the most serious mistake.

- **Simple math errors.** You may have incorrectly added or subtracted numbers, or used wrong formula or conversion factors.

- **Measurement errors.** You could have used the wrong scale for reduced-size drawings. For example, when drawings were half-size and you used the scale as shown, then your area was reduced by 75%, not the 50% reduction you thought to be.

- **Incorrect material prices.** You failed to get material price updates from the suppliers and the unit prices you used are too old.

- **Insufficient labor coverage.** You may have been too optimistic about certain items that will take longer to install than you allowed, or crews won't be available when the job starts.

- **Under estimated job duration.** You don't have enough money to cover jobsite overhead and risk paying for extra if not able to finish on time.

- **"Voluntary" price cuts.** You intentionally reduce your overhead or profits to get the job. The eagerness is not a valid excuse. Construction is a business that you should make reasonable profit on.

FIGURE 5.2 — TIPS TO REDUCE ESTIMATING MISTAKES

- Be organized and keep your desk clean
- Get information from complete set of drawings instead of a few sheets
- Read specs at least twice
- Use estimating checklists
- Use estimating forms
- Spend more time on large and expensive cost items
- Mark drawings when taking off items
- Prepare more detailed estimates instead of estimating by square feet
- Figure material and labor for each item instead of applying a combined unit rate or percentage
- Round up the results in each step of calculation and drop the pennies
- Have someone else check your estimate and take-off
- Check all formulas if using a spreadsheet program
- Compare costs with a similar project on a unit-price basis
- Always verify site conditions with the drawings
- Ask questions instead of making assumptions
- Take your time and never rush the estimate

FIGURE 5.3 — POST-BID ANALYSIS WORKSHEET

Job Name: _____

Estimate No.: _____ Estimator: _____

Type of Building: _____ Location: _____

Square Feet: _____ Functional Units: _____

Proposal Amount: $_____ Cost per SF: _____

Breakdown	% of Total Bid	Value per SF
Distribution Equipment		
Lighting Fixtures and Lamps		
Conduits and Fittings		
Wire and Cable		
Wiring Devices		
Electrical Heating		
Telephone and Data System		
Fire Alarm System		
Public Address System		
Security System		
Equipment Hook-up		
Electrical Site Services		
Demolition		
Concrete Work		
Excavation/Backfill		
Miscellaneous		
Jobsite Overhead		
Office Overhead		
Bond/Insurance		
Permit and License		
Allowance/Contingency		
Profit		
Total Bid		

FIGURE 5.4 — POST-BID DOCUMENT FOLDER

- A copy of bid proposal with fax confirmation, etc.
- Electrical drawings and specs
- A list of addendums
- Quantity take-off and pricing worksheets
- Site visit worksheet and photos
- Quotes from suppliers and subcontractors
- Bid correspondence with the customer
- Bond and insurance quote if applicable
- Estimating notes
- A list of questions asked and the answers

FIGURE 5.5 — QUESTION LIST FOR BID FOLLOW-UP

- Did the owner award the job?
- Did you receive all the information you requested?
- Do you have any questions on what you received?
- Is there anything else you need at this point?
- When would be a good time to check back with you?
- Who is the project manager, if such person has been assigned to the job?
- Who is the job superintendent?
- Who makes the decision regarding subcontract award?
- What are the important selection criteria?
- How important is price as a selection factor?
- What can we do to win your business?
- Could we schedule a meeting with you to discuss the proposal?
- If we did not get the job, who did?
- What was the price difference between our bid and winning price?

FIGURE 5.6 — COMMON VALUE ENGINEERING IDEAS

- Substitute load centers for panelboards

- Combine circuits into common conduit runs

- Reduce material quality (fixtures, equipment, etc.) to code grade standards

- Load lighting and plug circuits to the capacity of the circuit breaker protecting the wire

- Relocate panels to more central locations to cut down the lengths of conduit runs

- Reduce the size of panels and switchboards to the minimum capacity requirements

- Eliminate special installation details and instructions wherever possible

- Use aluminum wire instead of copper feeder wire if allowed

- Reduce wiring methods to standard code requirements

- Shift excavation, trenching, and backfill to others who can do it cheaper

- Shift concrete work to others who can do it cheaper

FIGURE 5.7 – BID HISTORY TRACKING WORKSHEET

Job Name	Bid Date	Job Size	Location	Job Type	Owner	General Contractor	Our Price	Bid Result

FIGURE 5.8 — POST-BID MARK-UP ANALYSIS

Job Name	Our Price	Our Cost	Our O & P	Awarded To	Competition's Price	Conclusion

Estimating Example:

Your bid was $50,000 and the winning bid was $42,000.

In your bid the total cost (material, labor, and subcontractor) was $ 40,000 and mark-up was $10,000 (i.e. 20%)

Was the competition charge too low on profit? Or did they incorrectly figure the job?

Possibility #1: You have the correct cost ($40,000). Now if the competition was also right, then their mark-up is only $ 2,000 (i.e. 4.7% of the total bid $42,000). It seems like they made a mistake.

Possibility #2: You found out you made an error and the cost should be $35,000 not $40,000. Now if the competition was right, then their mark-up is $7,000 (i.e. 16.7% of the total bid $42,000).

It normally takes quite a few jobs to find out the bidding pattern of your competition.

5-15

FIGURE 5.9 — PREPARE A REVISED ESTIMATE FOR CONTRACT

1. Request two complete sets of the latest drawings and specs (civil, architectural, structural, mechanical, electrical, etc.).

2. Review documents page by page to see what has changed since the bid. Write down the differences in detail.

3. Check the original list of inclusions, exclusions, and assumptions. See if there should be new issues added.

4. Track the original Request for Information (RFI) sent to the general contractor or design engineer. Send new RFI to ask more questions. Get all answers in writing (letter, fax, or e-mail). Only after an item has been clarified in writing to your satisfaction, can it be removed from the inclusion and exclusion list.

5. Perform revised quantity take-off with greater accuracy. If necessary, start a fresh one from scratch. Always keep the old and new estimates separate.

6. Get price updates from suppliers and subs. Make sure they have access to the latest design information. Prices might have changed dramatically since the bid.

7. Get a correct new total number and submit a revised proposal to the customer.

FIGURE 5.10 — ELEMENTS IN AN ELECTRICAL SUBCONTRACT

- Parties to the Contract

- Project Identification

- Scope of Work

- Schedule of Values

- Construction Schedule

- Submittals and Shop Drawings

- Quality

- Payment

- Change Orders

- Insurance, License, and Bond Requirements

- Warranty

- Protection of Work

- Safety

- Clean-up

- Termination

- Dispute Resolution

- General Clauses

- Attachments

FIGURE 5.11 — EXAMPLE WORK SCOPE IN AN ELECTRICAL SUBCONTRACT

The subcontractor (i.e. electrical contractor) should provide all labor materials, equipment, tools, and supervision required to furnish and install all electrical and appurtenances required for a complete installation in accordance with the contract documents, applicable codes, and governing agencies. The work includes, but is not limited to, the following:

1. Furnish and install all electrical fixtures, switches, meter bases, breaker boxes, and wiring.
2. Provide one meter base per unit and one house meter per building.
3. Furnish and install all exhaust fans. Ducting is installed by others.
4. Furnish, install, and provide power to the air-conditioner condenser.
5. Furnish, install, and provide power to the furnace electrical box.
6. Provide electrical hook-up of dishwasher and disposal.
7. Install range hood and microwave combination.
8. Install electrical clothes dryers.
9. Furnish and install wiring, jacks, boxes, and other items required to pre-wire these systems: telephone, cable television, security, intercom/sound, and home theater.

FIGURE 5.12 — NEGATIVE CLAUSES IN A BAD ELECTRICAL SUBCONTRACT

Clauses should be eliminated or revised:

- **"Pay if Paid".** Your payment from the general contractor depends on if they receive payment from the owner first.

- **"No Damage for Delay".** You will be allowed no more than a time extension for the delay you did not cause, and absolutely no monetary compensation.

- **"Unconditional Lien Waiver Before Payment".** You have to give up all of your lien and bond rights first, or they will not pay you.

- **"Change Order by Notice".** You must perform the additional work as directed by the general contractor without a written agreement in advance.

- **"Dispute Resolution by Litigation".** Any disputes will be resolved in the court of the general contractor's home state, which is far from where the project is located.

- **"Broad Form Indemnity".** You are required to name general contractors or owners as "additional names insured" on your insurance policy.

- **"Subcontract Supersedes Prior Negotiation".** This contract will overrule your bid proposal or any other prior negotiations.

- **"Acceptance of Final Payment as Waiver".** You give up all your legal rights to existing unresolved claims when you receive all payment in full as stated by this subcontract.

FIGURE 5.13 — "TURN-OVER" MEETING AGENDA

1. Have a job overview. Everyone should know the job name and location, building square footage, name of client and design engineer, etc.

2. Review contract documents, including drawings, specs, and addendums.

3. Review bid proposal and contract amount. What pricing assumptions were made?

4. Discuss site conditions and material storage problems.

5. Examine potential design problems, requests for information, and change orders.

6. Discuss project schedule and long lead items.

7. Review low subs, suppliers, and price escalation possibilities.

8. Evaluate risks. Make a list of make/break issues affecting the job costs.

FIGURE 5.14 — SIMPLIFIED JOB COST CODES	
Cost Code	Description
010	Lighting Fixtures and Lamps
020	Switchgear
030	Branch Conduit and Fitting
040	Feeder Conduit and Fittings
050	Wire Installation
060	Feeder Cable
070	Wiring Devices
080	Motor Control Equipment
090	Low-Voltage Systems Fit-Up
100	Equipment Fit-Up
200	Trenching and Excavation
300	Concrete Encasement
400	Manholes and Pre-cast Boxes
500	Demolition
600	Change Orders
700	Miscellaneous/Other

FIGURE 5.15 — CHANGE ORDER PRICING PROCEDURES

1. Review the total change request, including those sections that seem to be for other trades (i.e. mechanical).
2. Review original drawings and specs, and compare with revised documents.
3. Make a detailed take-off of all the material and work items.
4. Check the jobsite to verify the actual conditions, make notes, and take pictures if necessary.
5. Evaluate the affects of the change order on labor productivity.
6. Adjust the quantity take-off to reflect actual situation.
7. Complete the pricing and notify the general contractor.

NOTES:

- Do not handle a change without first receiving a written promise of payment for that change. Notify the general contractor if the change was requested directly by the owner.
- Do not deduct overhead and profit for deductive change orders.
- Remember to request additional time if the change order will delay your work.
- Do not forget, you almost always have to pay for cancellation, freight, and restocking charges of fixture and panelboards.
- Inform field workers of the deductive change so they can stop work in the area affected by that change.
- Keep records of change order requests and drawings.

FIGURE 5.16 — PHONE MEMO WORKSHEET

CONFIRMATION OF VERBAL COMMUNICATION

TO: DATE:

TIME:

ATTN:

PROJECT:

SUBJECT:

THIS MEMO CONFIRMS THE CONVERSATION:

BETWEEN   AND

ON (DATE)

BRIEF SUMMARY OF CONVERSATION:

PLEASE NOTIFY OUR OFFICE IF YOU DO NOT CONCUR WITH THIS
CONVERSATION SUMMARY.

FIGURE 5.17 — UNIT PRICING EXAMPLE

Electrical estimate for a school project using unit pricing method.

Items	Qty.	Unit	Unit Price	Subtotal
Service and Distribution				
Permit and set-up	1	l/s	$5,000.00	$5,000
Modifications to existing distribution	1	l/s	$3,000.00	$3,000
400 amp secondary panel	1	No	$3,000.00	$3,000
200 amp 120/208 V house panel	2	No	$1,500.00	$3,000
Motor control center	1	l/s	Excluded	$0
Feeder cables	3	No	$1,500.00	$4,500
Lighting				
Light fixtures: Ref 200 – pot-lights	17	No	$150.00	$2,550
Light fixtures: Ref 210 – susp. cylinders	41	No	$250.00	$10,250
Light fixtures: Ref 220 – downlights	15	No	$225.00	$3,375
Light fixtures: Ref 230 – spots on track	107	No	$95.00	$10,165
Light fixtures: Ref 240 – pot-lights	16	No	$200.00	$3,200
Light fixtures: Ref 250 – wall sconces	3	No	$250.00	$750
Light fixtures: Ref 260 – fluorescent	3	No	$125.00	$375
Fixture installation	219	No	$25.00	$5,475
Wiring (less spots on track)	96	No	$120.00	$11,520
Lnr. mtrs. of track	115	No	$75.00	$8,625
Track installation (including conduit and wiring)	33	No	$300.00	$9,900
Master relay panel and lighting control	1	No	$2,500.00	$2,500
Exit signs	1	No	$165.00	$165

FIGURE 5.17 — UNIT PRICING EXAMPLE *(cont.)*

Electrical estimate for a school project using unit pricing method.

Items	Qty.	Unit	Unit Price	Subtotal
Lighting *(cont.)*				
Lamps	1	No	Included	$0
Switches – single pole	1	No	$65.00	$65
Switches – double pole	3	No	$75.00	$225
IR/day-light sensors	1	No	Excluded	$0
Reposition existing light fixture	1	No	$100.00	$100
Power				
New duplex receptacle	28	No	$100.00	$2,800
Quad receptacle	2	No	$250.00	$500
Duplex receptacle GFI	3	No	$110.00	$330
Junction box	1	No	$100.00	$100
Miscellaneous power supplies	8	No	$350.00	$2,800
Mechanical power supplies	1	No	$1,500.00	$1,500
Fire Alarm				
Modifications to existing system	1	l/s	$4,500.00	$4,500
Telephone and Data				
New telephone, data outlets, and wiring	2	No	$450.00	$900
Pay telephone outlets	2	No	$350.00	$700
Comm. room equipment	1	No	$1,000.00	$1,000
TV/cable – allowance	4	No	$250.00	$1,000
Miscellaneous conduits	1	l/s	$1,000.00	$1,000
Security				
Security rough-in – allowance	1	l/s	$2,500.00	$2,500
Public Address				
Sound system rough-in – allowance	1	l/s	N/A	$0
Total Electrical Costs				**$109,100**

FIGURE 5.18 – UNIT PRICING WORKSHEET

Unit price to supply and install ONE standard duplex receptacle (commercial type).

Description	Quantity	Material			Labor		
		Unit Price	Per	Extension	Hours	Per	Extension
Receptacle	1	$ 0.50	E	$ 0.50	0.17	E	0.17
Plate	1	$ 0.19	E	$ 0.19	0.05	E	0.05
1-gg, 1/2" raise ring	1	$ 0.40	E	$ 0.40	0.05	E	0.05
4" Square box	1	$ 0.54	E	$ 0.54	0.17	E	0.17
1/2" EMT	20'	$13.00	C	$ 2.60	0.02	E	0.40
Connector	2	$ 0.09	E	$ 0.18	0.06	E	0.12
Coupling	1½	$ 0.09	E	$ 0.14	0.03	E	0.05
Strap with fastener	2½	$ 0.09	E	$ 0.23	0.03	E	0.08
Fasteners	3	$ 0.04	E	$ 0.12	0.01	E	0.03
#12 THHN wire	46'	$30.00	M	$ 1.38	0.01	E	0.12
Grounding pigtail	1	$ 0.22	E	$ 0.22	0.07	E	0.07
Wire nuts	2	$ 0.05	E	$ 0.10	–	–	–
Circuit breaker	1/6	$ 4.00	E	$ 0.67	0.12	E	0.02
Tapes, etc.	–	–	–	$ 0.11	–	–	0.01

Material	—	—	—	$ 7.38	—
Sales Tax	5%	—	—	$ 0.37	—
Material Subtotal	—	—	—	$ 7.75	—
Man-hours	—	—	—	—	1.34
Labor Hourly Rate	—	—	—	—	$ 35.00
Labor Subtotal	—	—	—	—	$ 46.90
Material and Labor Subtotal	$ 54.65	—	—	—	—
Allow 30% for Overhead and Profit	$ 16.39	—	—	—	—
Total Unit Price	**$71.04**	—	—	—	—

Notes: "E" stands for "Each". "C" stands for "Hundred", "M" stands for "Thousand".

5-27

FIGURE 5.19 — ESTIMATING "DESIGN-BUILD" JOBS

1. Determine the power requirements of the job. Draw a complete engineered riser layout with conduit and wire sizes, along with a voltage drop calculation.

2. Create a detailed list of all the equipment and apply a unit price to each.

3. Figure out the approximate quantity of fixtures and apply unit prices.

4. Quantify the service and feeders, and apply a per-foot unit price.

5. Check with your vendors for pricing on systems such as fire alarm and security, etc. The vendor may already have readily available, preliminary numbers which can be applied to this project.

6. Consult with the customer with respect to any special equipment requirements. Develop a needs list so you will have an understanding of the customers expectations and concept of the finished product.

FIGURE 5.20 — ELECTRICAL CONTRACT RELATIONSHIP

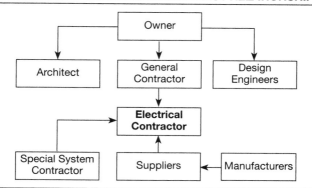

CHAPTER 6
Man-hour Tables

Disclaimer: These man-hour tables represent the author's best judgment and care for the information published. Instructions should always be carefully studied before using the data. Neither the author, DᴇWALT, nor the publisher is responsible for any losses or damages with respect to the accuracy, correctness, value, and sufficiency of the data contained herein.

There are many factors deciding how much work a man can do within one hour, such as jobsite conditions, supervision, tools and equipment, weather, code requirements, design quality, project location, etc.

The man-hour tables in this chapter are based on an average electrician working under normal conditions: new construction with fair productivity, standard materials and straight-forward installation, appropriate tools, and good coordination with other trades. You will always need to evaluate specific job conditions and make adjustments.

Most of the man-hours listed are for a single unit of certain item and not for assemblies. For example, unless noted otherwise, conduits are empty and separate from wires, connectors, couplings, straps, etc.

UNDERSTANDING MAN-HOUR TABLES

How accurate are the man-hour tables in this chapter? For example: Does it really take half an hour to install that junction box?

The man-hours listed in these tables include:

- Unloading, storing, and getting raw materials
- Getting and returning tools/equipment
- Normal time lost due to work breaks
- Reading drawings and discussing the work to be performed
- Normal handling, measuring, cutting, and fitting
- Field measurement and layout
- Test and balance of system
- Supervision time (foremen and super-intendants, etc.)
- Setting up and tearing down of lifts and ladders
- Regular clean-up of construction debris
- Infrequent correction or repairs required because of faulty installation

Please keep in mind when writing these tables, we don't know:

- Specific experience or training of your crew
- Plans or specs of your target job
- Where your job is and what local electrical code applies to your job

ADJUSTING STANDARD MAN-HOURS

Estimating is an art, not a science. Installation times for electrical items vary from job to job, from crew to crew, and even for the same crew, from day to day. This book is not a replacement for well-informed judgment. Use the following worksheet to adjust the standard man-hours in this chapter.

Working Conditions	Plus Percentage	Minus Percentage
Weather		
Crew Skills/Supervision		
Type of Work/Degree of Difficulty		
Size of Job/Economy of Scale		
Site Congestion		
Inspections/Specs		
Working Hours/Overtime		
Distance to Stocking Pile		
Assembled/Unassembled Material		
Mounting Heights		
Work Space		
Total Adjustment Percentage		

Estimating Example:

It normally takes 3.00 hours to set a small transformer. But now you are estimating a difficult job and decide 25% needs to be added to everything to cover possible productivity loss due to site conditions.

Then the adjusted man-hour for that transformer should be 3.00 × (100% + 25%) = 3.75 hours for one transformer installation.

WRITING YOUR OWN MAN-HOUR TABLES

If you feel more comfortable in writing your own man-hour tables, the next few pages will help you do it properly.

Estimating Math:

Total Man-hours = Number of Working Crew Members × Hours Crew Worked

Unit Man-hour for the Item = Crew Output/ Total Man-hours

Estimating Example:

An electrician and his helper spent one 8 hour day to install 200 feet of conduits

Total man-hours are: 2 people × 8 hours = 16 hours

One linear foot of such conduit will take:
16 hours/200 feet = 0.08 hours/foot

When recording man-hour information, you can specify the following:

1. Materials and installation methods (i.e. dimensions and types of conduits)
2. Type of jobs (i.e. new or renovation, residential, or non-residential)
3. Crew (i.e. just one person or a team including a supervision and apprentice)
4. Tools and equipment (i.e. scaffolding, lift, etc.)
5. Weather conditions (i.e. rain, snow, wind, and temperature)

Check your job at the same time each morning and record the number of units (such as the length of conduits, fixtures, etc.) installed the previous day. Repeat these on different projects over a period of time to verify your result.

DAILY JOBSITE MAN-HOUR WORKSHEET

Job Name: _____ Ref No.: _____

Superintendent: _____

Date: _____ Worksheet Number: _____

Item	Daily Output	Crew Member	Crew Hours	Total Man-hours	Man-hour Per Unit

TIPS FOR ESTIMATING AND IMPROVING LABOR PRODUCTIVITY

Labor is always the hardest part to estimate, even harder to control for job costing.

For estimators:

- Study plans and specs carefully
- Know what your men can handle
- Get familiar with tool and equipment inventory
- Divide unfamiliar tasks into smaller components to estimate
- Keep good records of actual field performance

For project managers:

- Work out a realistic schedule and stick to it
- Have submittals approved in a timely manner
- Get right materials on site, in the right quantities, at the right time
- Make tools and equipment available on the job, when needed
- Have drawings accurately marked up, including all documentation
- Track job progress and crew productivity properly
- Train better workers through company supported education
- Have the home office provide timely administrative support to the field
- Improve communications with general contractor and all other trades
- Have a clean, safe, and "drug-free" site to work in

CONVERTING MINUTES TO DECIMAL HOURS

Minutes	Decimal in Hours	Minutes	Decimal in Hours
1	0.017	31	0.517
2	0.033	32	0.533
3	0.050	33	0.550
4	0.067	34	0.567
5	0.083	35	0.583
6	0.100	36	0.600
7	0.117	37	0.617
8	0.133	38	0.633
9	0.150	39	0.650
10	0.167	40	0.667
11	0.183	41	0.683
12	0.200	42	0.700
13	0.217	43	0.717
14	0.233	44	0.733
15	0.250	45	0.750
16	0.267	46	0.767
17	0.283	47	0.783
18	0.300	48	0.800
19	0.317	49	0.817
20	0.333	50	0.833
21	0.350	51	0.850
22	0.367	52	0.867
23	0.383	53	0.883
24	0.400	54	0.900
25	0.417	55	0.917
26	0.433	56	0.933
27	0.450	57	0.950
28	0.467	58	0.967
29	0.483	59	0.983
30	0.500	60	1.000

LIGHTING FIXTURES

INCANDESCENT LIGHT FIXTURE

Item	Hours per Fixture
Surface mounted	0.20
Cover mounted	0.25
Ceiling, 2 lamps	0.30
Ceiling, 3 lamps	0.35
Ceiling, 4 lamps	0.40
Ceiling, with canopy	0.45
Wall, 1 lamp	0.40
Wall, 2 lamps	0.50
Recessed	0.60

TRACK LIGHT FIXTURE

Item	Hours per Fixture
2' starter	0.30
4' starter	0.40
8' starter	0.55
Lamp holder	0.20
Decorator	0.20

EXIT LIGHT FIXTURE

Item	Hours per Fixture
Single face	0.50
Double face	0.55
With rechargeable batteries	0.75

LIGHTING FIXTURES *(cont.)*	
FLUORESCENT LIGHT FIXTURE	
Item/Size	**Hours per Fixture**
18", one lamp	0.25
18", two lamps	0.30
24", one lamp	0.35
24", two lamps	0.45
36", one lamp	0.40
36", two lamps	0.55
48", one lamp	0.45
48", two lamps	0.60
72", one lamp	0.50
72", two lamps	0.70
96", one lamp	0.75
96", two lamps	0.90

LIGHTING FIXTURES (cont.)

T-BAR FLUORESCENT LIGHTING FIXTURE

Item/Size	Hours per Fixture
12", one lamp	0.55
12", two lamps	0.60
12", three lamps	0.65
20", two lamps	0.65
20", three lamps	0.70
20", four lamps	0.75
24", two lamps	0.70
24", three lamps	0.75
24", four lamps	0.80
48", four lamps	1.25
48", six lamps	1.30
48", eight lamps	1.35

LIGHTING FIXTURES *(cont.)*	
DISCHARGE LIGHTING FIXTURE	
Item/Size	**Hours per Fixture**
10" diameter	0.50
16" diameter	0.70
18" diameter	0.70
29" diameter	0.75
24" square	0.60
HID EXTERIOR FLOODLIGHTS	
Item/Size	**Hours per Fixture**
Bracket-mounted	1.30
Pole-mounted	0.70
Wall-mounted	0.80
HID ROADWAY LUMINARIES	
Item/Size	**Hours per Fixture**
28" housing	0.80
29" housing	0.85
30" housing	1.00
40" housing	1.50
CEILING FANS WITH LIGHTS	
Size	**Hours per Fixture**
36"	1.45
42"	1.70
44"	1.70
52"	1.70
60"	1.80

LIGHTING POLES

STEEL LIGHT POLES

Height	Hours per Pole
8'	0.80
10'	0.85
12'	0.95
14'	1.10
16'	1.30
18'	1.60
20'	1.70
24'	1.80
30'	2.25
35'	3.00
39'	4.00
50'	7.00
60'	8.00

ALUMINUM LIGHT POLES

Height	Hours per Pole
6'	0.60
8'	0.70
10'	0.75
12'	0.85
15'	0.90
18'	1.15
20'	1.25
25'	1.35
30'	1.50
35'	1.90
40'	2.50
45'	3.00
50'	3.75

LAMPS	
Item	**Hours per Lamp**
Incandescent	0.04
Discharge	0.10
Sodium	0.15
Fluorescent	0.06
Special	0.10

OUTLET BOXES	
STEEL BOXES AND PARTS	
Box/Parts	**Hours per Item**
4" and 4 11/16"	0.15
Extension rings	0.07
Blanks	0.06
Plaster rings	0.05
Surface covers	0.07
Bar hangers	0.03

MASONRY BOXES	
Type	**Hours per Box**
1 gg	0.20
2 gg	0.30
3 gg	0.40
4 gg	0.50
5 gg	0.60
6 gg	0.70
7 gg	0.80
8 gg	0.90

CAST BOXES

Type	Hours per Box
1/2"	0.30
3/4"	0.32
1"	0.35

JUNCTION BOXES

Type	Hours per Box
$4 \times W \times 4$	0.15
$6 \times W \times 4$	0.15
$8 \times W \times 4$	0.20
$10 \times W \times 4$	0.25
$12 \times W \times 4$	0.40
$6 \times W \times 6$	0.20
$8 \times W \times 6$	0.25
$10 \times W \times 6$	0.40
$12 \times W \times 6$	0.60
$15 \times W \times 6$	0.90
$18 \times W \times 6$	1.00
$24 \times W \times 6$	1.40

SWITCHES	
Item	**Hours per Switch**
1-pole	0.20
3-way	0.25
2-pole	0.25
4-way	0.30
Dimmer	0.25
Combination	0.35
Double throw, 1-pole	0.30
Double throw, 2-pole	0.35
Time switch	1.00
Swimming pool timer	1.25
Calendar dial	1.50
RECEPTACLES	
Item	**Hours per Receptacle**
15 amp	0.25
20 amp	0.30
30 amp	0.35
50 amp	0.40
60 amp	0.50
COVER PLATES	
Item	**Hours per Plate**
1 gang	0.05
2 gang	0.10
3 gang	0.15
4 gang	0.20
5 gang	0.25
6 gang	0.30

SWITCHBOARDS	
Size	Hours per Switchboard
800 amp	10.00
1200 amp	16.00
1600 amp	20.00
2000 amp	26.00
3000 amp	39.00
4000 amp	52.00
DISTRIBUTION PANELS	
Size	Hours per Panel
400 amp	6.00
600 amp	8.00
800 amp	10.00
1200 amp	16.00
1600 amp	20.00
2000 amp	26.00
MOTOR CONTROL CENTERS	
Size	Hours per Center
400 amp	6.00
600 amp	8.00
800 amp	10.00
1200 amp	16.00
1600 amp	20.00
2000 amp	26.00

CIRCUIT BREAKERS

600 V, 2-POLE

Size	Hours per Breaker
60 amp	0.30
100 amp	0.70
225 amp	1.00
400 amp	1.50
600 amp	3.00
800 amp	3.00
1000 amp	5.00
1200 amp	5.00

600 V, 3-POLE

Size	Hours per Breaker
60 amp	0.34
100 amp	0.75
225 amp	1.25
400 amp	1.80
600 amp	3.60
800 amp	3.60
1000 amp	5.40
1200 amp	5.40

120/240 V

Size	Hours per Breaker
1-pole	0.20
2-pole	0.40
3-pole	0.45

270/480 V

Size	Hours per Breaker
1-pole	0.30
2-pole	0.50
3-pole	0.70

PANELBOARDS	
100 AMP	
Size	**Hours per Panelboard**
8-pole	3.40
10-pole	4.00
12-pole	4.60
14-pole	5.70
16-pole	6.30
18-pole	6.90
20-pole	7.50
200 AMP	
Size	**Hours per Panelboard**
22-pole	8.10
24-pole	8.70
26-pole	9.80
28-pole	10.40
30-pole	11.00
32-pole	11.60
34-pole	12.20
36-pole	12.80
38-pole	13.90
40-pole	14.50
42-pole	15.10
400 AMP	
Size	**Hours per Panelboard**
38-pole	14.40
40-pole	15.00
42-pole	15.60

LOAD CENTERS

Size/Type	Hours per Load Center
100 amp MLO, 20 space	3.80
100 amp MCB, 20 space	4.25
200 amp MLO, 40 space	5.50
200 amp MCB, 40 space	6.00

METERS

Size/Type	Hours per Meter
1 phase, 200 amp	0.70
3 phase, 200 amp	0.90
Per additional (meter paks)	0.50
Current transformer cabinet	2.40

HUBS

Size	Hours per Hub
3/4"	0.20
1"	0.25
1 1/4"	0.25
1 1/2"	0.30
2"	0.30
2 1/2"	0.40
Closure	0.10

TRANSFORMERS — DRY TYPE	
SMALL TRANSFORMERS	
Size	**Hours per Transformer**
0.10 kVA	0.25
0.15 kVA	0.30
0.25 kVA	0.30
0.50 kVA	0.35
0.75 kVA	0.40
1.0 kVA	0.50
1.5 kVA	0.55
2.0 kVA	0.60
3.0 kVA	0.75
5.0 kVA	1.00
6.0 kVA	1.00
7.5 kVA	1.25
9.0 kVA	1.40
10.0 kVA	1.50
LARGE TRANSFORMERS	
Size	**Hours per Transformer**
15 kVA	3.20
30 kVA	6.00
45 kVA	10.00
75 kVA	12.00
112.5 kVA	14.00
150 kVA	16.00
225 kVA	20.00
300 kVA	24.00
500 kVA	28.00

TRANSFORMERS VIBRATION ISOLATORS	
Size	**Hours per Isolator**
15 kVA	0.70
30 kVA	0.70
45 kVA	0.90
75 kVA	0.90
112.5 kVA	1.00
150 kVA	1.00
225 kVA	1.20
300 kVA	1.50
500 kVA	1.50
EMERGENCY GENERATOR	
Size	**Hours per Generator**
10.0 kw	10.00
15.0 kw	12.00
30.0 kw	18.00
50.0 kw	22.00
75.0 kw	24.00
100.0 kw	30.00
125.0 kw	36.00
150.0 kw	38.00
175.0 kw	40.00
200.0 kw	50.00
250.0 kw	54.00
300.0 kw	62.00
350.0 kw	74.00
400.0 kw	84.00
500.0 kw	98.00

SAFETY SWITCHES — NEMA 1, 12, AND 3R

2-POLE

Size	Hours per Switch
30 amp	0.60
60 amp	0.75
100 amp	1.00
200 amp	1.50
400 amp	3.50
600 amp	5.00
800 amp	7.50
1200 amp	12.00

3-POLE

Size	Hours per Switch
30 amp	0.70
60 amp	0.90
100 amp	1.20
200 amp	1.80
400 amp	4.00
600 amp	6.00
800 amp	8.50
1200 amp	13.00

3-POLE WITH SN

Size	Hours per Switch
30 amp	0.80
60 amp	1.00
100 amp	1.40
200 amp	2.00
400 amp	4.50
600 amp	6.50
800 amp	9.00
1200 amp	13.50

SAFETY SWITCHES — NEMA 4 AND 5	
2-POLE	
Size	**Hours per Switch**
30 amp	1.50
60 amp	2.00
100 amp	2.50
200 amp	3.00
400 amp	5.80
600 amp	11.50
3-POLE	
Size	**Hours per Switch**
30 amp	1.80
60 amp	2.20
100 amp	2.70
200 amp	3.50
400 amp	7.00
600 amp	13.00
FUSES	
Items	**Hours per Fuse**
30-200 amp	0.05
200-400 amp	0.08
400-600 amp	0.11
600-800 amp	0.20
800-1200 amp	0.30
RELAYS	
Items	**Hours per Relay**
2-pole	0.70
3-pole	0.90
4-pole	1.10
6-pole	1.50
10-pole	2.25
Relay mounting track, 18"	0.20
Relay mounting track, 36"	0.30

MOTOR STARTERS	
MANUAL STARTERS	
Size	Hours per Starter
1-pole	0.60
2-pole	0.70
3-pole, 3 hp	1.25
3-pole, 5 hp	1.25
3-pole, 7.5 hp	1.50
3-pole, 10 hp	1.50
MAGNETIC STARTERS	
Size	Hours per Starter
00	1.50
0	1.70
1	1.80
2	2.50
3	4.00
4	5.00
5	6.00
6	8.00
7	12.00
COMBO STARTERS	
Size	Hours per Starter
0	2.25
1	2.70
2	3.25
3	5.50

SIGNAL CABINETS	
Size	Hours per Cabinet
12" × 12" × 4"	0.45
12" × 16" × 4"	0.60
12" × 16" × 6"	0.70
12" × 18" × 4"	0.90
12" × 18" × 6"	1.00
12" × 24" × 4"	1.20
12" × 24" × 6"	1.30
18" × 18" × 4"	1.20
18" × 18" × 6"	1.30
18" × 24" × 4"	1.50
18" × 24" × 6"	1.60
18" × 30" × 4"	1.80
18" × 30" × 6"	1.90
24" × 24" × 4"	2.10
24" × 24" × 6"	2.20
24" × 30" × 4"	2.50
24" × 30" × 6"	2.60
24" × 36" × 4"	3.00
24" × 36" × 6"	3.10
30" × 30" × 4"	3.00
30" × 30" × 6"	3.10
30" × 36" × 4"	3.30
30" × 36" × 6"	3.40
36" × 48" × 4"	3.60
36" × 48" × 6"	3.70

FEEDER BUS DUCTS

COPPER FEEDER BUS DUCT

Size	Hours per Foot
800 amp	0.25
1000 amp	0.30
1200 amp	0.35
1350 amp	0.40
1600 amp	0.45
2000 amp	0.50
2500 amp	0.55
3000 amp	0.60
4000 amp	0.80
5000 amp	1.00

ALUMINUM FEEDER BUS DUCT

Size	Hours per Foot
600-1000 amp	0.20
1200-1600 amp	0.25
2000 amp	0.30
2500 amp	0.40
3000 amp	0.50
4000 amp	0.60

PLUG-IN BUS DUCTS	
COPPER PLUG-IN BUS DUCT	
Size	**Hours per Foot**
225 amp	0.20
400 amp	0.25
600 amp	0.30
800 amp	0.35
1000 amp	0.40
1200 amp	0.45
1350 amp	0.50
1600 amp	0.55
2000 amp	0.60
2500 amp	0.70
3000 amp	0.90
ALUMINUM PLUG-IN BUS DUCT	
Size	**Hours per Foot**
225-600 amp	0.20
800-1000 amp	0.25
1200 amp	0.30
1350 amp	0.35
1600 amp	0.45
2000 amp	0.50
2500 amp	0.55
3000 amp	0.60
4000 amp	0.70

BUS DUCT ELBOWS

COPPER BUS DUCT ELBOWS

Size	Hours per Elbow
225-600 amp	1.20
800-1600 amp	1.35
2000-3000 amp	1.45
4000-5000 amp	1.70

ALUMINUM BUS DUCT ELBOWS

Size	Hours per Elbow
225-600 amp	1.00
800-1600 amp	1.20
2000-3000 amp	1.30
4000 amp	1.40

BUS DUCT END CLOSURES

COPPER BUS DUCT END CLOSURES

Size	Hours per End Closure
225-600 amp	0.40
800-1600 amp	0.50
2000-3000 amp	0.60
4000-5000 amp	0.80

ALUMINUM BUS DUCT END CLOSURES

Size	Hours per End Closure
225-600 amp	0.30
800-1600 amp	0.40
2000-3000 amp	0.50
4000 amp	0.70

BUS DUCT TAP BOXES	
COPPER BUS DUCT TAP BOXES	
Size	**Hours per Tap Box**
225 amp	2.20
400 amp	2.80
600 amp	4.20
800 amp	4.40
1000 amp	5.70
1200 amp	5.90
1350 amp	7.20
1600 amp	7.40
2000 amp	9.00
2500 amp	12.00
3000 amp	15.00
4000 amp	19.00
5000 amp	24.00
ALUMINUM BUS DUCT TAP BOXES	
Size	**Hours per Tap Box**
225 amp	2.00
400 amp	2.60
600 amp	4.00
800 amp	4.50
1000 amp	5.50
1200 amp	6.00
1350 amp	7.00
1600 amp	7.50
2000 amp	9.00
2500 amp	11.50
3000 amp	14.50
4000 amp	19.00

BUS DUCT CIRCUIT BREAKER ADAPTORS

COPPER BUS DUCT CIRCUIT BREAKER ADAPTORS

Size	Hours per Adaptor
225-600 amp	1.20
800-1600 amp	1.40
2000 amp	1.50

ALUMINUM BUS DUCT CIRCUIT BREAKER ADAPTORS

Size	Hours per Adaptor
225-600 amp	1.00
800-1600 amp	1.20
2000 amp	1.30

TIME CLOCKS

Items	Hours per Time Clock
1-pole, 24 hr.	0.60
3-pole, 24 hr.	0.80
3-pole, 7 day	0.94
3-pole, 7 day with reserve	1.10
Astronomic dial with reserve	1.20

PHOTO CELLS

Ceiling Height	Hours per Cell
Mounted at 12 ft.	0.45
Mounted at 20 ft.	0.70

GRS CONDUITS AND FITTINGS		

RIGID

GRS CONDUITS	
Size	**Hours per Foot**
1/2"	0.04
3/4"	0.05
1"	0.05
11/4"	0.07
11/2"	0.08
2"	0.10
21/2"	0.12
3"	0.14
31/2"	0.17
4"	0.20
5"	0.25
6"	0.33

GRS FITTINGS		
Size	**Hours per Elbow**	**Hours per Coupling**
1/2"	0.20	0.15
3/4"	0.25	0.19
1"	0.30	0.24
11/4"	0.38	0.27
11/2"	0.40	0.29
2"	0.45	0.32
21/2"	0.62	0.50
3"	0.85	0.60
31/2"	0.98	0.70
4"	1.20	0.80
5"	1.60	0.90
6"	2.00	1.00

IMC CONDUITS AND FITTINGS

IMC CONDUITS

Size	Hours per Foot
1/2"	0.04
3/4"	0.04
1"	0.05
11/4"	0.07
11/2"	0.08
2"	0.09
21/2"	0.11
3"	0.13
31/2"	0.15
4"	0.17

IMC FITTINGS

Size	Hours per Elbow	Hours per Coupling
1/2"	0.20	0.05
3/4"	0.25	0.06
1"	0.30	0.08
11/4"	0.38	0.10
11/2"	0.40	0.10
2"	0.44	0.15
21/2"	0.59	0.15
3"	0.79	0.20
31/2"	0.89	0.20
4"	1.00	0.25

EMT CONDUITS AND FITTINGS

EMT CONDUITS

Size	Hours per Foot
1/2"	0.04
3/4"	0.04
1"	0.05
1 1/4"	0.07
1 1/2"	0.08
2"	0.09
2 1/2"	0.11
3"	0.13
3 1/2"	0.15
4"	0.17

EMT ELBOWS

Size	Hours per Elbow
1"	0.10
1 1/4"	0.12
1 1/2"	0.15
2"	0.17
2 1/2"	0.18
3"	0.20
3 1/2"	0.22
4"	0.24

EMT STRAPS

Size	Hours per Strap
1/2"	0.03
3/4"	0.04
1"	0.05
1 1/4"	0.06
1 1/2"	0.06
2"	0.10
2 1/2"	0.10
3"	0.15
3 1/2"	0.15
4"	0.15

EMT CONDUITS AND FITTINGS (cont.)

EMT CONNECTOR

Size	Set Screw Type	Compression Type
1/2"	0.06	0.08
3/4"	0.06	0.08
1"	0.08	0.09
11/4"	0.10	0.10
11/2"	0.10	0.12
2"	0.15	0.15
21/2"	0.18	0.23
3"	0.20	0.25
31/2"	0.22	0.27
4"	0.25	0.30

EMT COUPLINGS

Size	Set Screw Type	Compression Type
1/2"	0.03	0.05
3/4"	0.03	0.05
1"	0.04	0.06
11/4"	0.05	0.07
11/2"	0.07	0.09
2"	0.10	0.12
21/2"	0.15	0.20
3"	0.17	0.22
31/2"	0.19	0.24
4"	0.21	0.27

PVC CONDUITS AND FITTINGS

PVC CONDUITS

Size	Hours per Foot	Size	Hours per Foot	Size	Hours per Foot
1/2"	0.03	11/2"	0.06	31/2"	0.11
3/4"	0.03	2"	0.07	4"	0.12
1"	0.04	21/2"	0.09	5"	0.13
11/4"	0.05	3"	0.10	6"	0.14

PVC FITTINGS

Size	Hours per Elbow	Hours per Coupling	Hours per Adaptor	Hours per End Bell	Hours per Cap/Plug
1/2"	0.10	0.05	0.10	0.10	0.05
3/4"	0.10	0.05	0.10	0.10	0.06
1"	0.13	0.05	0.13	0.10	0.08
11/4"	0.15	0.06	0.15	0.15	0.10
11/2"	0.18	0.08	0.16	0.15	0.10
2"	0.20	0.10	0.18	0.20	0.10
21/2"	0.25	0.10	0.20	0.20	0.10
3"	0.30	0.15	0.25	0.25	0.15
31/2"	0.35	0.15	0.30	0.30	0.15
4"	0.40	0.15	0.35	0.30	0.15
5"	0.50	0.20	0.40	0.35	0.20
6"	0.60	0.25	0.40	0.40	0.20

PVC-COATED STEEL CONDUITS AND FITTINGS

PVC-COATED STEEL CONDUITS

Size	Hours per Foot	Size	Hours per Foot
1/2"	0.05	21/2"	0.15
3/4"	0.06	3"	0.17
1"	0.07	31/2"	0.19
11/4"	0.09	4"	0.21
11/2"	0.11	5"	0.25
2"	0.13	—	—

PVC-COATED STEEL FITTINGS

Size	Hours per Elbow	Hours per Coupling	Hours per Cut/Thread
1/2"	0.10	0.05	0.25
3/4"	0.10	0.06	0.25
1"	0.15	0.08	0.30
11/4"	0.20	0.10	0.40
11/2"	0.20	0.10	0.40
2"	0.25	0.15	0.50
21/2"	0.30	0.20	0.75
3"	0.35	0.20	1.00
31/2"	0.40	0.25	1.25
4"	0.50	0.25	1.30
5"	0.75	0.30	1.50

ALUMINUM RIGID CONDUITS AND FITTINGS		
ARC CONDUITS		
Size		**Hours per Foot**
1/2"		0.04
3/4"		0.05
1"		0.05
11/4"		0.06
11/2"		0.07
2"		0.09
21/2"		0.10
3"		0.12
31/2"		0.14
4"		0.16
5"		0.20
6"		0.25
ARC FITTINGS		
Size	**Hours per Elbow**	**Hours per Coupling**
1/2"	0.15	0.05
3/4"	0.20	0.06
1"	0.25	0.07
11/4"	0.30	0.08
11/2"	0.32	0.09
2"	0.37	0.10
21/2"	0.50	0.11
3"	0.60	0.12
31/2"	0.70	0.13
4"	0.80	0.14
5"	1.00	0.15
6"	1.50	0.16

FLEX CONDUITS AND FITTINGS

FLEX CONDUITS

Size	Hours per Foot
3/8"	0.03
1/2"	0.03
3/4"	0.03
1"	0.03
1 1/4"	0.04
1 1/2"	0.04
2"	0.04
2 1/2"	0.04
3"	0.05
3 1/2"	0.05
4"	0.05

FLEX FITTINGS

Size	Hours per Fitting
3/8"	0.05
1/2"	0.05
3/4"	0.06
1"	0.08
1 1/4"	0.10
1 1/2"	0.15
2"	0.20
2 1/2"	0.25
3"	0.25
3 1/2"	0.30
4"	0.30

LIQUID-TIGHT FLEX CONDUITS AND FITTINGS

LIQUID-TIGHT FLEX CONDUITS

Size	Hours per Foot
3/8"	0.04
1/2"	0.04
3/4"	0.05
1"	0.05
1 1/4"	0.06
1 1/2"	0.07
2"	0.09
2 1/2"	0.11
3"	0.15
4"	0.17

LIQUID-TIGHT FLEX FITTINGS

Size	Hours per Fitting
3/8"	0.15
1/2"	0.15
3/4"	0.15
1"	0.20
1 1/4"	0.25
1 1/2"	0.25
2"	0.30
2 1/2"	0.30
3"	0.40
4"	0.40

ENT CONDUITS AND FITTINGS

ENT CONDUITS		FLEX FITTINGS	
Size	**Hours per Foot**	**Size**	**Hours per Fitting**
1/2"	0.02	1/2"	0.03
3/4"	0.02	3/4"	0.04
1"	0.03	1"	0.05

NIPPLES	
Size	**Hours per Nipple**
1/2"	0.10
3/4"	0.10
1"	0.10
11/4"	0.15
11/2"	0.15
2"	0.15
21/2"	0.20
3"	0.20
31/2"	0.20
4"	0.25
5"	0.40
6"	0.60

CHASE NIPPLES	
Size	**Hours per Nipple**
1/2"	0.05
3/4"	0.06
1"	0.07
11/4"	0.08
11/2"	0.09
2"	0.10
21/2"	0.12
3"	0.15
31/2"	0.17
4"	0.20
5"	0.25
6"	0.30

NIPPLES *(cont.)*	
OFFSET NIPPLES	
Size	**Hours per Nipple**
1/2"	0.10
3/4"	0.15
1"	0.20
11/4"	0.25
11/2"	0.25
2"	0.30
21/2"	0.35
3"	0.35
31/2"	0.40
4"	0.45
SERVICE HEAD-CLAMP TYPE	
Size	**Hours per Service Head**
1/2"	0.30
3/4"	0.32
1"	0.37
11/4"	0.50
11/2"	0.55
2"	0.63
21/2"	0.70
3"	0.85
31/2"	1.25
4"	2.10

CONDULETS	
TYPE LB, LL, LR, C, AND E	
Size	**Hours per Condulet**
1/2"	0.20
3/4"	0.25
1"	0.30
1 1/4"	0.35
1 1/2"	0.37
2"	0.45
2 1/2"	0.60
3"	0.75
3 1/2"	0.90
4"	1.20
TYPE T AND X	
Size	**Hours per Condulet**
1/2"	0.30
3/4"	0.38
1"	0.45
1 1/4"	0.53
1 1/2"	0.56
2"	0.68
2 1/2"	0.90
3"	1.13
3 1/2"	1.35
4"	1.80

CONDULET COVERS	
Size	Hours per Cover
1/2"	0.10
3/4"	0.10
1"	0.10
1 1/4"	0.12
1 1/2"	0.12
2"	0.12
2 1/2"	0.15
3"	0.15
3 1/2"	0.20
4"	0.20
SEAL-OFF FITTINGS	
Size	Hours per Fitting
1/2"	0.50
3/4"	0.60
1"	0.65
1 1/4"	0.70
1 1/2"	0.75
2"	0.90
2 1/2"	1.40
3"	1.65
4"	2.00

EXPANSION FITTINGS

4" CAST EXPANSION FITTINGS	
Size	**Hours per Fitting**
1/2"	0.40
3/4"	0.45
1"	0.60
11/4"	0.70
11/2"	0.80
2"	0.90
21/2"	1.10
3"	1.10
31/2"	1.30
4"	1.50
8" CAST EXPANSION FITTINGS	
Size	**Hours per Fitting**
1/2"	0.45
3/4"	0.55
1"	0.75
11/4"	0.85
11/2"	0.95
2"	1.05
21/2"	1.10
3"	1.45
31/2"	1.85
4"	2.10

3-PIECE COUPLINGS	
Size	**Hours per Coupling**
1/2"	0.20
3/4"	0.22
1"	0.25
1 1/4"	0.30
1 1/2"	0.35
2"	0.40
2 1/2"	0.60
3"	0.75
3 1/2"	0.90
4"	1.25
5"	1.65
6"	2.10

BUSHINGS	
Size	**Hours per Bushing**
1/2"	0.10
3/4"	0.10
1"	0.10
11/4"	0.15
11/2"	0.15
2"	0.20
21/2"	0.20
3"	0.25
31/2"	0.25
4"	0.30
5"	0.40
6"	0.50
INSULATED BUSHINGS	
Size	**Hours per Bushing**
1/2"	0.10
3/4"	0.12
1"	0.15
11/4"	0.20
11/2"	0.22
2"	0.25
21/2"	0.50
3"	0.65
31/2"	0.75
4"	0.85
5"	1.05
6"	1.35

LOCKNUTS	
Size	**Hours per Locknut**
1/2"	0.05
3/4"	0.06
1"	0.08
11/4"	0.10
11/2"	0.10
2"	0.20
21/2"	0.20
3"	0.20
31/2"	0.25
4"	0.30
5"	0.35
6"	0.40

CONDUIT STRAPS	
Size	**Hours per Strap**
1/2"	0.05
3/4"	0.06
1"	0.80 .08
11/4"	0.10
11/2"	0.10
2"	0.10
21/2"	0.15
3"	0.20
31/2"	0.20
4"	0.25
5"	0.25
6"	0.30

CLAMP BACK	
Size	Hours per Clamp
1/2"	0.05
3/4"	0.06
1"	0.08
1 1/4"	0.10
1 1/2"	0.10
2"	0.10
2 1/2"	0.15
3"	0.15
3 1/2"	0.20
4"	0.20
5"	0.25
6"	0.25
ENCLOSURES	
Size	Hours per Enclosure
16 × 12 × 6	0.50
24 × 20 × 6	0.70
30 × 24 × 6	0.80
20 × 20 × 8	0.80
30 × 20 × 8	0.90
36 × 24 × 8	1.00
42 × 30 × 8	1.50
60 × 36 × 8	2.50
60 × 36 × 10	3.00
Plywood backing	0.15

WIREWAY AND FITTINGS

WIREWAY

Size	Hours per Foot
2¹/₂" × 2¹/₂"	0.05
4" × 4"	0.06
6" × 6"	0.07
8" × 8"	0.10
10" × 10"	0.12
12" × 12"	0.15

WIREWAY FITTINGS

Size	Hours per Elbow	Hours per Coupling	Hours per Flange	Hours per Tee
2¹/₂" × 2¹/₂"	0.20	0.10	0.20	0.25
4" × 4"	0.20	0.10	0.20	0.25
6" × 6"	0.20	0.10	0.30	0.30
8" × 8"	0.30	0.15	0.40	0.40
10" × 10"	0.40	0.20	0.50	0.50
12" × 12"	0.50	0.25	0.70	0.70

UNDERFLOOR DUCTS AND FITTINGS

UNDERFLOOR RACEWAY

Type	Hours per Foot
Duct Type #1	0.03
Duct Type #2	0.04

UNDERFLOOR RACEWAY JUNCTION BOX

Type	Hours per Box
4 Type #1 Ducts	0.50
8 Type #1 Ducts	0.60
12 Type #1 Ducts	0.75
4 Type #2 Ducts	0.60
8 Type #2 Ducts	1.00
4 Type #1 and 4 Type #2 Ducts	0.80

UNDERFLOOR RACEWAY SUPPORT

Type	Hours per Support
1 Type #1 Ducts	0.15
2 Type #1 Ducts	0.20
3 Type #1 Ducts	0.25
1 Type #2 Ducts	0.20
2 Type #2 Ducts	0.25
1 Type #1 and 1 Type #2 Ducts	0.25
2 Type #1 and 1 Type #2 Ducts	0.25
1 Type #1 and 2 Type #2 Ducts	0.30

UNDERFLOOR DUCTS AND FITTINGS *(cont.)*

#1 DUCT FITTINGS

Type	Hours per Fitting
Coupling	0.05
Elbow	0.25
Duct Plug	0.05
Cabinet Connector	0.30
Wye Connector	0.25

#2 DUCT FITTINGS

Type	Hours per Fitting
Coupling	0.06
Elbow	0.30
Duct Plug	0.06
Cabinet Connector	0.50
Wye Connector	0.30

SURFACE METAL RACEWAY AND FITTINGS

SURFACE METAL RACEWAY	
Size	**Hours per Foot**
1/2"	0.03
3/4"	0.04
1"	0.05
1 1/4"	0.06
2 3/4"	0.07
4 3/4"	0.08

SURFACE METAL RACEWAY FITTINGS	
Type	**Hours per Fitting**
Tee	0.20
Elbow	0.25
Adapter	0.20
Coupling	0.10
End fitting	0.10
Outlet box	0.25
Utility box	0.25
Extension box	0.25
Fixture box	0.30
Device box	0.40

CABLE TRAY AND FITTINGS

GALVANIZED CABLE TRAY

Size	Hours per Foot
6"	0.05
9"	0.08
12"	0.11
18"	0.13
24"	0.15
36"	0.17

ALUMINUM CABLE TRAY

Size	Hours per Foot
12"	0.084
18"	0.105
24"	0.125

CABLE TRAY FITTINGS

Item	Hours per Fitting
45 degree elbow	0.75
90 degree elbow	1.00
Tee	1.25
Adapter	0.30
End cap	0.20
Hanger	0.50

COPPER WIRES	
TYPES THHN, THW, XHHW, ETC.	
Size	Hours per Foot
No. 14 AWG	0.006
12	0.007
10	0.008
8	0.009
6	0.010
4	0.012
3	0.013
2	0.013
1	0.014
1/0	0.018
2/0	0.020
3/0	0.025
4/0	0.028
250 KCMIL	0.029
300	0.032
350	0.039
400	0.046
500	0.053
600	0.060
750	0.067
1000	0.077
No. 18 AWG	0.005
No. 16	0.006

COPPER WIRES *(cont.*	
5,000 V	
Size	**Hours per Foot**
No. 8 AWG	0.012
6	0.014
4	0.016
2	0.018
1	0.019
1/0	0.023
2/0	0.028
3/0	0.030
4/0	0.033
250 KCMIL	0.038
350	0.048
500	0.065
15,000 V	
Type	**Hours per Fitting**
No. 4 AWG	0.018
2	0.023
1	0.028
1/0	0.030
2/0	0.033
3/0	0.038
4/0	0.043
250 KCMIL	0.050
350	0.061
500	0.078

ALUMINUM WIRES — TYPES THHN, THW, XHHW, ETC.	
Size	Hours per Foot
No. 6 AWG	0.009
4	0.010
3	0.011
2	0.012
1	0.013
1/0	0.015
2/0	0.018
3/0	0.021
4/0	0.023
250 KCMIL	0.025
300	0.027
350	0.031
400	0.037
500	0.043
600	0.050
750	0.056
1000	0.065

FLEXIBLE CORDS	
Size	**Hours per Foot**
#18-2	0.008
#16-2	0.009
#14-2	0.010
#12-2	0.012
#10-2	0.014
#18-3	0.009
#16-3	0.010
#14-3	0.011
#12-3	0.013
#10-3	0.015
LOW VOLTAGE CABLES	
Size/Type	**Hours per Foot**
Coaxial cable	0.005
Thermostat cable	
4-conductor	0.005
8-conductor	0.010
Control cable	
4-conductor	0.005
8-conductor	0.009
12-conductor	0.013
20-conductor	0.023
30-conductor	0.025
50-conductor	0.031
Telephone cables	
2-pair	0.005
6-pair	0.010
10-pair	0.018
25-pair	0.032

BUILDING WIRE CABLES	
Size/Type	Hours per Foot
Type NM No. 14 or No. 12	
2-conductor with ground	0.006
3-conductor with ground	0.008
Type NM No. 10	
2-conductor with ground	0.008
Type NM No. 6	
2-conductor with ground	0.011
Type SEU	
No. 4	0.016
3	0.018
2	0.020
1	0.024
1/0	0.027
2/0	0.033
3/0	0.040
4/0	0.047
Type AC (BX) cables	
No. 14, 2-conductor	0.012
No. 14, 3-conductor	0.014
No. 14, 4-conductor	0.016
No. 12, 2-conductor	0.016
No. 12, 3-conductor	0.018
No. 12, 4-conductor	0.021
No. 10, 2-conductor	0.018
No. 10, 3-conductor	0.021

FLAT CONDUCTOR CABLES AND PARTS	
FLAT CONDUCTOR CABLES	
Size/Type	**Hours per Foot**
3-conductor cable	0.04
4-conductor cable	0.05
5-conductor cable	0.06
CABLE PARTS	
Size/Type	**Hours per Item**
Duplex receptacle service head	0.60
Double duplex receptacle service head	0.70
Telephone service head	0.60
Receptacle and telephone service head	0.70
Blank cover service head	0.05
Surface transition box	0.40
Recessed transition box	0.60
End caps	0.05
Insulators	0.15
Splice connectors	0.30
Tap connectors	0.30
Cable connectors	0.30
Terminal blocks	0.40

COMMUNICATION CABLES	
Type	Hours per Cable
1 to 4 pairs	0.006
5 to 8 pairs	0.007
9 to 12 pairs	0.008
13 to 16 pairs	0.009
17 to 20 pairs	0.010
25 pairs	0.012
30 pairs	0.013
40 pairs	0.015
50 pairs	0.018
60 pairs	0.020
RESIDENTIAL ITEMS/DEVICES	
Items	Hours per Item
Nail-on box	0.05
Ceiling box	0.07
Plastic finish plate	0.04
Weatherproof plate	0.06
Receptacle	0.06
GFI Receptacle	0.12
Flush dryer outlet	0.12
1-pole switch	0.05
3-way switch	0.06
Residential light fixture	0.25
Recessed fixture	0.40
Exhaust fan	0.30
Smoke detector, 120 V	0.20
Surface range outlet	0.20
Door chime	0.20
Push button	0.07

RESIDENTIAL (ROMEX) CABLES	
Size	**Hours per Foot**
#14-2	0.006
#14-3	0.007
#14-4	0.008
#12-2	0.006
#12-3	0.007
#12-4	0.009
#10-2	0.007
#10-3	0.008
#8-2	0.009
#8-3	0.010
#6-2	0.010
#6-3	0.011
#6-2 Aluminum	0.008
#4-2	0.011
#4-3	0.012
#2-2	0.012
#2-3	0.013
#1-2	0.013
#1-3	0.014

EQUIPMENT HOOK-UP	
KITCHEN EQUIPMENT	
Item	**Hours per Item**
Dishwasher	5.00
Food mixer	1.25
Food warmer	1.00
Fridge	1.50
Garbage disposal	1.25
Grill hood fan	2.00
Hot tables	2.00
Oven	1.50
Walk-in cooler/freezer	8.00
Water heater	2.50
SMALL MECHANICAL EQUIPMENT	
Item	**Hours per Item**
Air handler	2.00
Boiler	8.00
Chiller	24.00
Conveyor	16.00
Control panel	4.00
Cooling tower	10.00
Elevator	24.00
Exhaust fan	1.50
Furnace	2.50
Overhead door	6.00
Pump	2.00
Switch	1.00
Unit heater	2.25
Valve	1.50

HEATING DEVICES	
Item	**Hours per Device**
Baseboard heaters	
500-750 W	0.50
800-1250 W	0.65
1300-2000 W	0.75
2100-2500 W	1.00
Thermostat	0.40
Corner and blank section	0.20
Receptacle section	0.40
Wall heaters	
750-2000 W	1.00
2500-4000 W	1.40
Infrared heaters	
600-2000 W	0.70
2500-4000 W	1.00
Radiant ceiling panels	0.50
Radiant ceiling panels frame	0.20
SIGNALING SYSTEMS	
Item	**Hours per Item**
Bell	0.45
Beacon	0.50
Buzzer	0.30
Chime	0.40
Door opener	0.25
Horn	0.40
Push button	0.30
Siren	0.50
Transformer	0.30

PRE-CAST CONCRETE PULL BOX/MANHOLES

Sizes	Hours per Item
24" × 36" × 24"	1.00
24" × 36" × 36"	1.25
30" × 48" × 36"	1.50
36" × 72" × 24"	2.25
36" × 60" × 36"	2.50
48" × 48" × 36"	2.50
48" × 60" × 48"	2.75
48" × 48" × 72"	3.00
48" × 48" × 84"	3.00
48" × 78" × 86"	3.50
48" × 90" × 84"	3.50
54" × 102" × 78"	4.00
60" × 120" × 72"	4.50
60" × 120" × 86"	4.50
72" × 72" × 60"	3.75
72" × 72" × 72"	4.00
72" × 120" × 84"	4.50
72" × 180" × 108"	6.00
72" × 180" × 144"	6.50
72" × 228" × 108"	9.00
84" × 264" × 96"	12.00
96" × 120" × 72"	5.00
96" × 168" × 100"	8.00
96" × 168" × 112"	10.00
96" × 240" × 112"	12.00
96" × 312" × 112"	14.00
120" × 180" × 120"	16.00

PRE-CAST CONCRETE TRANSFORMER SLAB	
Sizes	**Hours per Item**
31" × 46"	0.50
32" × 38"	0.50
42" × 49"	0.70
44" × 46"	0.70
48" × 54"	1.00
57" × 76"	1.25
72" × 66"	1.50
73" × 75"	1.50
94" × 72"	1.50
MOTORS	
Sizes	**Hours per Motor**
Less than 1 hp	0.50
1 to 3 hp	0.75
5 to 15 hp	1.25
20 hp	1.50
30 hp	2.00
50 hp	3.00
75 hp	4.00
100 hp	5.00

TRENCH EXCAVATION				
HAND EXCAVATION				
Trench Dimension		Hours per Foot of Trench		
Depth	Width	Sandy Soil	Medium Soil	Clay Soil
1' 0"	18"	0.07	0.11	0.16
1' 6"	18"	0.11	0.17	0.24
2' 0"	18"	0.14	0.22	0.32
2' 6"	18"	0.18	0.28	0.40
3' 0"	18"	0.21	0.34	0.48
3' 6"	18"	0.25	0.40	0.56
4' 0"	18"	0.28	0.44	0.64
4' 6"	18"	0.32	0.50	0.72
5' 0"	24"	0.48	0.76	1.05
5' 6"	24"	0.53	0.84	1.16
6' 0"	24"	0.57	0.90	1.24
6' 6"	24"	0.62	1.00	1.38
7' 0"	24"	0.91	1.30	2.08
7' 6"	24"	1.00	1.43	2.28
8' 0"	24"	1.06	1.50	2.40
8' 6"	24"	1.13	1.60	2.56
9' 0"	24"	1.20	1.70	2.72
9' 6"	24"	1.27	1.80	2.88
10' 0"	24"	1.34	1.88	3.00
10' 6"	24"	1.41	1.95	3.12
11' 0"	24"	1.48	2.08	3.32
11' 6"	24"	1.56	2.18	3.48
12' 0"	24"	1.63	2.28	3.64

MACHINE EXCAVATION				
Trench Dimension		Hours per Foot of Trench		
Depth	Width	Sandy Soil	Medium Soil	Clay Soil
1' 0"	18"	0.01	0.01	0.01
1' 6"	18"	0.01	0.01	0.02
2' 0"	18"	0.01	0.01	0.02
2' 6"	18"	0.02	0.02	0.03
3' 0"	18"	0.02	0.02	0.03
3' 6"	18"	0.02	0.02	0.03
4' 0"	18"	0.03	0.03	0.04
4' 6"	18"	0.03	0.03	0.04
5' 0"	24"	0.05	0.05	0.06
5' 6"	24"	0.05	0.05	0.06
6' 0"	24"	0.06	0.06	0.07
6' 6"	24"	0.06	0.06	0.07
7' 0"	24"	0.07	0.07	0.08
7' 6"	24"	0.07	0.07	0.08
8' 0"	24"	0.08	0.08	0.09
8' 6"	24"	0.08	0.08	0.09
9' 0"	24"	0.09	0.09	0.10
9' 6"	24"	0.09	0.09	0.10
10' 0"	24"	0.10	0.10	0.11
10' 6"	24"	0.10	0.10	0.11
11' 0"	24"	0.11	0.11	0.12
11' 6"	24"	0.11	0.12	0.13
12' 0"	24"	0.12	0.12	0.13

TRENCH BACKFILL AND GRADING			
Trench Dimension		**Hours per Foot of Trench**	
Depth	**Width**	**Backfill**	**Grading**
1' 0"	18"	0.03	0.03
1' 6"	18"	0.04	0.03
2' 0"	18"	0.06	0.03
2' 6"	18"	0.07	0.04
3' 0"	18"	0.09	0.04
3' 6"	18"	0.10	0.04
4' 0"	18"	0.11	0.04
4' 6"	18"	0.13	0.04
5' 0"	24"	0.19	0.04
5' 6"	24"	0.21	0.04
6' 0"	24"	0.23	0.04
6' 6"	24"	0.25	0.06
7' 0"	24"	0.26	0.06
7' 6"	24"	0.29	0.06
8' 0"	24"	0.30	0.06
8' 6"	24"	0.32	0.06
9' 0"	24"	0.34	0.06
9' 6"	24"	0.36	0.08
10' 0"	24"	0.38	0.08
10' 6"	24"	0.39	0.08
11' 0"	24"	0.42	0.08
11' 6"	24"	0.44	0.08
12' 0"	24"	0.46	0.08

PIT EXCAVATION	
Size	**Hours per Pit**
24" × 24" × 24"	0.14
24" × 36" × 24"	0.18
24" × 48" × 24"	0.20
24" × 24" × 36"	0.21
24" × 36" × 36"	0.27
24" × 48" × 36"	0.30
24" × 24" × 48"	0.28
24" × 36" × 48"	0.36
24" × 48" × 48"	0.40

CONCRETE CUTTING	
Cut Depth	**Hours per Foot of Saw Cut**
1"	0.03
11/2"	0.03
2"	0.04
21/2"	0.04
3	0.05
31/2"	0.05
4"	0.06
5"	0.07

CHANNELING

CHANNELING REINFORCED CONCRETE

Size	Hours per Foot
1/2"	0.35
3/4"	0.38
1"	0.40
11/4"	0.45
11/2"	0.50
2"	0.55
21/2"	0.65
3	0.70
31/2"	0.75
4"	0.90

CHANNELING BRICK

Size	Hours per Foot
1/2"	0.32
3/4"	0.35
1"	0.38
11/4"	0.40
11/2"	0.43
2"	0.44
21/2"	0.50
3	0.52
31/2"	0.57
4"	0.60

CUTTING HOLES IN MASONRY

Size	Hours per Foot
3/4" × 8"	0.40
3/4" × 12"	0.50
3/4" × 16"	0.60
3/4" × 20"	0.70
3/4" × 24"	0.80
1 1/2" × 8"	0.60
1 1/2" × 12"	0.70
1 1/2" × 16"	0.85
1 1/2" × 20"	0.95
1 1/2" × 24"	1.10
3" × 8"	1.30
3" × 12"	1.50
3" × 16"	1.70
3" × 20"	1.90
3" × 24"	2.00
5" × 8"	1.85
5" × 12"	2.00
5" × 16"	2.20
5" × 20"	2.50
5" × 24"	2.80

CORE DRILLING	
Hole Size	**Hours per Hole**
3/4" × 8"	0.60
3/4" × 12"	0.80
3/4" × 16"	1.00
3/4" × 20"	1.20
3/4" × 24"	1.40
1 1/2" × 8"	0.80
1 1/2" × 12"	1.00
1 1/2" × 16"	1.30
1 1/2" × 20"	1.50
1 1/2" × 24"	1.80
3" × 8"	1.70
3" × 12"	2.10
3" × 16"	2.50
3" × 20"	2.80
3" × 24"	3.30
5" × 8"	2.30
5" × 12"	2.70
5" × 16"	3.20
5" × 20"	3.80
5" × 24"	4.20
SLEEVES	
Type	**Hours per Sleeve**
Wall	0.80
Floor	0.40

CHAPTER 7
Estimating Forms

Standard forms help in the estimating process. Using forms or worksheets is a good way to prepare logical and consistent estimates, however, they do not guarantee that everything needed to be estimated will be covered. Contract documents and job conditions should always be carefully evaluated first before trying to use any estimating forms.

In this chapter, some estimating forms are grouped for your convenience in quantity take-off and pricing. These blank quantity take-off forms can be photocopied for repeated use, or even customized to suit your special needs, such as computerized estimating programs.

If you have any questions regarding how estimating forms work, what to take-off, or how to convert quantities to prices, please refer to previous chapters for more information.

FIGURE 7.1 — BID LOAD SCHEDULE WORKSHEET

Estimate #	Job Name	Job Size	Bid Date	Estimator

FIGURE 7.2 — GENERAL JOB INFORMATION WORKSHEET

Job Name: _____ Estimate No.: _____

Estimator: _____

Bid Due Date: _____ Time: _____

Job Location: _____ Tax Rate: _____

Construction Type: _____ New Construction _____ Renovation

Gross Floor Area: _____ Number of Floors: _____

No. of: Condo Units____ Hospital Beds ____ Hotel Rooms ____

Send Bid To: _____ Owner _____ General Contractor

Owner: _____ Tel. No: _____

General Contractor: _____ Tel. No: _____

Architect: _____ Tel. No. _____

Electrical Engineer: _____ Tel. No. _____

Plans/Specs or Design/Build: _____ CAD required? ____

Job Start Date: _____ Job Finish Date: _____ Phases: ___

Retainage: _____ Liquidated Damages: _____

Warranty Period: _____

Labor Conditions: ___Prevailing Wages __ Union __ Open Shop

Bond Required: __Yes __ No

Pricing Requirement: ____ Firm Price ____ Negotiated

Send RFI To: _____ General Contractor _____ Electrical Engineer
          _____ Other

Work Done by Owner's Contractor: _____

Cost Breakdown Requirements: _____

Allowances: _____

Alternates: _____

7-3

FIGURE 7.3 – BUILDING AREA WORKSHEET

Floor Levels	Floor Area	Perimeter	Floor Height	Ceiling Height	Structure Details	Units/ Rooms
Basement						
1st Floor						
2nd Floor						
3rd Floor						
4th Floor						
Total						

FIGURE 7.4 — GENERAL QUANTITY TAKE-OFF WORKSHEET

Job Name: _____

Estimate No.: _____ Estimator: _____

Date: _____ Worksheet Page No.: _____

Item Description	Details				Unit of Measure	Waste %	Total Qty.
	Pieces	Length	Width	Height			
Total							

FIGURE 7.5 — LIGHTING FIXTURES
TAKE-OFF WORKSHEET

Job Name: _____

Estimate No.: _____ Estimator: _____

Date: _____ Worksheet Page No.: _____

Fixture Type	Basement	1st Floor	2nd Floor	Subtotal

FIGURE 7.6 — LAMPS TAKE-OFF WORKSHEET

Job Name: _____

Estimate No.: _____ Estimator: _____

Date: _____ Worksheet Page No.: _____

Fixture Type	Fixture Quantity	Lamp Type	Lamp Type	Lamp Type	Lamp Type
Total					

FIGURE 7.7 — CONDUIT TAKE-OFF WORKSHEET

Job Name: _____

Estimate No.: _____ Estimator: _____

Date: _____ Worksheet Page No.: _____

Conduit Type/Size	Basement	1st Floor	2nd Floor	Subtotal
Total				

FIGURE 7.8 — FITTING TAKE-OFF WORKSHEET

Job Name: _____

Estimate No.: _____   Estimator: _____

Date: _____   Worksheet Page No.: _____

Fitting Type/Size	Basement	1st Floor	2nd Floor	Subtotal
Total				

FIGURE 7.9 — WIRE TAKE-OFF WORKSHEET

Job Name: _____

Estimate No.: _____ Estimator: _____

Date: _____ Worksheet Page No.: _____

Conduit Type/Size	Conduit Length	Wire Size	Wire Size	Wire Size	Wire Size
Total					

FIGURE 7.10 — BOXES AND DEVICES
TAKE-OFF WORKSHEET

Job Name: _____

Estimate No.: _____   Estimator: _____

Date: _____   Worksheet Page No.: _____

Devices/ Boxes	Basement	1st Floor	2nd Floor	3rd Floor	Subtotal

FIGURE 7.11 — EQUIPMENT/PANELS TAKE-OFF WORKSHEET

Job Name: _____

Estimate No.: _____ Estimator: _____

Date: _____ Worksheet Page No.: _____

Equipment/Panel	Description	Quantity

FIGURE 7.12 – EQUIPMENT/PANEL FEEDER SCHEDULE WORKSHEET

Routing				Raceway						Trench	Wire			
From	To	Amps	Size	Type	Length	90 ELL	Term	Strap			Number	Size	Volts	Length

7-13

FIGURE 7.13 — MATERIAL PRICING WORKSHEET

Job Name: _____

Estimate No.: _____ Estimator: _____

Date: _____ Worksheet Page No.: _____

Item	Quantity	Unit Price	Per	Extension
Subtotal				
Sales Tax				
Freight				
Total Material Costs				

Notes about column "Per":
E: each
C: Hundred
M: Thousand
Lot or L/S: Lump Sum
SF: Square Foot
CY: Cubic Yard

FIGURE 7.14 – SUPPLIER EVALUATION WORKSHEET

Item	Quantity	Supplier #1		Supplier #2		Supplier #3	
		Unit Price	Extension	Unit Price	Extension	Unit Price	Extension
Subtotal							
Sales Tax							
Freight							
Total							

FIGURE 7.15 — LABOR HOURLY RATE WORKSHEET	
Names of Field Employees	**Total Annual Wages**
Subtotal	
Add:	
Bonus	
Living Allowances	
Social Security	
Medicare	
Federal Unemployment Tax (FUTA)	
State Unemployment Tax (SUTA)	
Worker's Compensation Insurance	
General Liability Insurance	
Health Insurance	
Dental Insurance	
Pension (401K)	
Union Dues	
Total Annual Payroll	
Total Working Hours	
Vacations	
Sick Leave	
Total Hours Paid	
Average Labor Hourly Rate	

FIGURE 7.16 — LABOR PRICING WORKSHEET

Job Name: _____

Estimate No.: _____ Estimator: _____

Date: _____ Worksheet Page No.: _____

Item	Quantity	Man-hour	Per	Extension
Total Man-hours				
Labor Hourly Rate				
Burden				
Labor Cost Subtotal				

Notes about column "Per":
E: each
C: Hundred
M: Thousand
Lot or L/S: Lump Sum
SF: Square Foot
CY: Cubic Yard

FIGURE 7.17 — SUBCONTRACTOR EVALUATION WORKSHEET

System Quoted	Plug Quantity/ Price	Sub #1	Sub #2	Sub #3
Base Price				
Work Included				
Delivery Time				
Adjustments				
Sales Tax				
Delivery to Jobsite				
Complete Installation				
Per Plans and Specs				
Exclusions				
Addenda Received				
Other Factors				
Adjusted Total Bid				

FIGURE 7.18 — PRICING SUMMARY WORKSHEET

Item	Rate	Amount	Subtotal
Material Quotes			
Switchgear			
Fixtures and lamps			
Conduit and fittings			
Boxes and wires			
Devices			
Hook-up			
Miscellaneous			
Sales Tax			
Material Subtotal			
Labor Subtotal			
Subcontractor Subtotal			
Jobsite Overhead			
Total Direct Costs			
Office overhead			
Owner's allowance			
Bond/insurance			
Electrical permit			
Contingency			
Profit			
Total Bid Price			

FIGURE 7.19 – UNIT PRICING FORMS FOR RESIDENTIAL CONSTRUCTION

DUPLEX RECEPTACLE

Description	Quantity	Material Unit Price	Per	Extension	Labor Hours	Per	Extension
Plastic nail-on box	1						
Duplex receptacle	1						
Plastic plate	1						
#14 cable	20'						
Circuit breaker	1/6						
Staples, wire nuts, etc.	1						
Material							
Sales Tax							
Material Subtotal							
Man-hours							
Labor Hourly Rate							
Labor Subtotal							
Material and Labor Subtotal							
Allow 50% for Overhead and Profit							
Total Unit Price							

FIGURE 7.19 – UNIT PRICING FORMS FOR RESIDENTIAL CONSTRUCTION (cont.)

GROUND-FAULT INTERRUPTER RECEPTACLE

Description	Quantity	Material			Labor		
		Unit Price	Per	Extension	Hours	Per	Extension
Plastic nail-on box	1						
GFI receptacle	1						
Finish plate	1						
#14 cable	20'						
Circuit breaker	1/6						
Staples, wire nuts, etc.	1						
Material							
Sales Tax							
Material Subtotal							
Man-hours							
Labor Hourly Rate							
Labor Subtotal							
Material and Labor Subtotal							
Allow 50% for Overhead and Profit							
Total Unit Price							

FIGURE 7.19 – UNIT PRICING FORMS FOR RESIDENTIAL CONSTRUCTION (cont.)

20 AMP OUTLET (WASHER, FREEZER, ETC.)

Description	Quantity	Material			Labor		
		Unit Price	Per	Extension	Hours	Per	Extension
Plastic box	1						
Receptacle	1						
Finish plate	1						
#12/2 cable	24'						
Circuit breaker	1						
Staples, etc.							
Material							
Sales Tax							
Material Subtotal							
Man-hours							
Labor Hourly Rate							
Labor Subtotal							
Material and Labor Subtotal							
Allow 50% for Overhead and Profit							
Total Unit Price							

FIGURE 7.19 — UNIT PRICING FORMS FOR RESIDENTIAL CONSTRUCTION (cont.)

DRYER OUTLET

| Description | Quantity | Material | | | Labor | | |
		Unit Price	Per	Extension	Hours	Per	Extension
Plastic box	1						
Dryer receptacle	1						
Finish plate	1						
#10/2 cable	24'						
2-pole circuit breaker	1						
Staples, etc.							
Material							
Sales Tax							
Material Subtotal							
Man-hours							
Labor Hourly Rate							
Labor Subtotal							
Material and Labor Subtotal							
Allow 50% for Overhead and Profit							
Total Unit Price							

7-23

FIGURE 7.19 – UNIT PRICING FORMS FOR RESIDENTIAL CONSTRUCTION (cont.)

WATER HEATER

Description	Quantity	Material Unit Price	Per	Extension	Labor Hours	Per	Extension
#10/2 cable	24'						
2-pole circuit breaker	1						
Cable connector	1						
Terminations	3						
Staples, etc.							
Material							
Sales Tax							
Material Subtotal							
Man-hours							
Labor Hourly Rate							
Labor Subtotal							
Material and Labor Subtotal							
Allow 50% for Overhead and Profit							
Total Unit Price							

FIGURE 7.19 – UNIT PRICING FORMS FOR RESIDENTIAL CONSTRUCTION (cont.)

AIR CONDITIONING UNIT

Description	Quantity	Material Unit Price	Material Per	Material Extension	Labor Hours	Labor Per	Labor Extension
#10/2 cable	30'						
Disconnect switch–WP	1						
1/2" Sealtite	4'						
Connectors – Straight	2						
#10 THHN wire	18'						
2-pole circuit breaker	1						
Terminations	9						
Staples, wire nuts, etc.							
Material							
Sales Tax							
Material Subtotal							
Man-hours							
Labor Hourly Rate							
Labor Subtotal							
Material and Labor Subtotal							
Allow 50% for Overhead and Profit							
Total Unit Price							

7-25

FIGURE 7.19 – UNIT PRICING FORMS FOR RESIDENTIAL CONSTRUCTION (cont.)

AIR HANDLING UNIT/HEATER

Description	Quantity	Material			Labor		
		Unit Price	Per	Extension	Hours	Per	Extension
#6/2 cable	22'						
Disconnector switch	1						
Cable connector	3						
2-pole circuit breaker	1						
Terminations	9						
Staples, screws, etc.							
Material							
Sales Tax							
Material Subtotal							
Man-hours							
Labor Hourly Rate							
Labor Subtotal							
Material and Labor Subtotal							
Allow 50% for Overhead and Profit							
Total Unit Price							

FIGURE 7.19 – UNIT PRICING FORMS FOR RESIDENTIAL CONSTRUCTION *(cont.)*

RANGE OUTLETS

Description	Quantity	Material			Labor		
		Unit Price	Per	Extension	Hours	Per	Extension
Surface receptacle	1						
#6/2 cable	24'						
2-pole circuit breaker	1						
Staples, screws, etc.							
Material							
Sales Tax							
Material Subtotal							
Man-hours							
Labor Hourly Rate							
Labor Subtotal							
Material and Labor Subtotal							
Allow 50% for Overhead and Profit							
Total Unit Price							

FIGURE 7.19 – UNIT PRICING FORMS FOR RESIDENTIAL CONSTRUCTION (cont.)

SINGLE-POLE SWITCH

Description	Quantity	Material Unit Price	Per	Extension	Hours	Labor Per	Extension
Plastic box	1						
1-pole switch	1						
Finish plate	1						
#14 Romex	15'						
Staples, etc.							
Material							
Sales Tax							
Material Subtotal							
Man-hours							
Labor Hourly Rate							
Labor Subtotal							
Material and Labor Subtotal							
Allow 50% for Overhead and Profit							
Total Unit Price							

FIGURE 7.19 – UNIT PRICING FORMS FOR RESIDENTIAL CONSTRUCTION *(cont.)*

LIGHT FIXTURE OUTLET

Description	Quantity	Material			Labor		
		Unit Price	Per	Extension	Hours	Per	Extension
#14 Romex	15'						
Round plastic box	1						
Circuit breaker	1/6						
Fixture and lamp	1						
Staples, wire nuts, etc.							
Material							
Sales Tax							
Material Subtotal							
Man-hours							
Labor Hourly Rate							
Labor Subtotal							
Material and Labor Subtotal							
Allow 50% for Overhead and Profit							
Total Unit Price							

FIGURE 7.19 — UNIT PRICING FORMS FOR RESIDENTIAL CONSTRUCTION *(cont.)*

WELL PUMP

Description	Quantity	Material Unit Price	Material Per	Material Extension	Labor Hours	Labor Per	Labor Extension
#14 Romex	20'						
#14 UF cable	50'						
Trenching	42'						
WP box	1						
2-pole switch	1						
WP switch cover	1						
1/2" PVC conduit	5'						
90 ELL	1						
TA	1						
1/2" Sealtite	5'						
Connectors – Straight	1						
Strap	1						
Terminations	3						
2-pole circuit breaker	1						
Wire nuts, etc.							

Material																		
Sales Tax																		
Material Subtotal																		
Man-hours																		
Labor Hourly Rate																		
Labor Subtotal																		
Material and Labor Subtotal																		
Allow 50% for Overhead and Profit																		
Total Unit Price																		

FIGURE 7.19 – UNIT PRICING FORMS FOR RESIDENTIAL CONSTRUCTION *(cont.)*

BATH EXHAUST FAN

| Description | Quantity | Material | | | Labor | | |
		Unit Price	Per	Extension	Hours	Per	Extension
Exhaust fan	1						
#14 Romex	16'						
Terminations	3						
Circuit breaker	1/10						
Cable connector	1						
Staples, etc.							
Material							
Sales Tax							
Material Subtotal							
Man-hours							
Labor Hourly Rate							
Labor Subtotal							
Material and Labor Subtotal							
Allow 50% for Overhead and Profit							
Total Unit Price							

FIGURE 7.19 – UNIT PRICING FORMS FOR RESIDENTIAL CONSTRUCTION (cont.)

DOUBLE FLOODLIGHT

| Description | Quantity | Material | | | Labor | | |
		Unit Price	Per	Extension	Hours	Per	Extension
#14 Romex	30'						
Box	1						
Cluster base	1						
Lamp holder	2						
PAR lamp	2						
Circuit breaker	1/4						
Staples, wire nuts, etc.							
Material							
Sales Tax							
Material Subtotal							
Man-hours							
Labor Hourly Rate							
Labor Subtotal							
Material and Labor Subtotal							
Allow 50% for Overhead and Profit							
Total Unit Price							

FIGURE 7.19 – UNIT PRICING FORMS FOR RESIDENTIAL CONSTRUCTION (cont.)

TELEPHONE OUTLET

Description	Quantity	Material			Labor		
		Unit Price	Per	Extension	Hours	Per	Extension
Plastic box	1						
Finish plate	1						
Phone cable	32'						
Staples, etc.							
Material							
Sales Tax							
Material Subtotal							
Man-hours							
Labor Hourly Rate							
Labor Subtotal							
Material and Labor Subtotal							
Allow 50% for Overhead and Profit							
Total Unit Price							

FIGURE 7.19 – UNIT PRICING FORMS FOR RESIDENTIAL CONSTRUCTION (cont.)

DOOR CHIME AND PUSH BUTTON

| Description | Quantity | Material | | | Labor | | |
		Unit Price	Per	Extension	Hours	Per	Extension
Chime and transformer	1						
Push button	1						
Phone cable	25'						
2-gan box	1						
#14 cable	15'						
Circuit breaker	1/10						
Staples, wire nuts, etc.							
Material							
Sales Tax							
Material Subtotal							
Man-hours							
Labor Hourly Rate							
Labor Subtotal							
Material and Labor Subtotal							
Allow 50% for Overhead and Profit							
Total Unit Price							

FIGURE 7.19 – UNIT PRICING FORMS FOR RESIDENTIAL CONSTRUCTION (cont.)

150 AMP OVERHEAD SERVICE

Description	Quantity	Material			Labor		
		Unit Price	Per	Extension	Hours	Per	Extension
1 1/2" conduit riser	10'						
1 1/2" service head	1						
1 1/2" connector/termination	1						
1 1/2" meter hub	1						
1 1/2" straps	2						
#2/0 aluminum THW wire	30'						
#1 aluminum THW wire	15'						
Meter (furnished by utility)	1						
#6 bare copper wire	30'						
Pipe clamp	1						
Ground rod	1						
Ground rod clamp	1						
1/2" EMT	10'						
1/2" EMT connector	1						
1/2" EMT strap	1						
Main circuit breaker	1						
#2/0 aluminum SEU cable	28'						
Large cable connectors	2						
Tapes, etc.							

Material																				
Sales Tax																				
Material Subtotal																				
Man-hours																				
Labor Hourly Rate																				
Labor Subtotal																				
Material and Labor Subtotal																				
Allow 50% for Overhead and Profit																				
Total Unit Price																				

FIGURE 7.20 – UNIT PRICING FORMS FOR COMMERCIAL CONSTRUCTION

DUPLEX RECEPTACLE

Description	Quantity	Material Unit Price	Per	Material Extension	Labor Hours	Per	Labor Extension
Receptacle	1						
Plate	1						
1-gg. 1/2" raise ring	1						
4" square box	1						
1/2" EMT	20'						
Connector	2						
Coupling	1½						
Strap and fastener	2½						
Fasteners	3						
#12 THHN wire	46'						
Grounding pigtail	1						
Wire nuts	2						
Circuit breaker	1/6						
Tapes, etc.							

													Material
													Sales Tax
													Material Subtotal
													Man-hours
													Labor Hourly Rate
													Labor Subtotal
													Material and Labor Subtotal
													Allow 30% for Overhead and Profit
													Total Unit Price

7-39

FIGURE 7.20 — UNIT PRICING FORMS FOR COMMERCIAL CONSTRUCTION *(cont.)*

SINGLE-POLE SWITCH

Description	Quantity	Material			Labor		
		Unit Price	Per	Extension	Hours	Per	Extension
1-pole switch	1						
Plate	1						
1-gg, 1/2" raise ring	1						
4" square box	1						
1/2" EMT	16'						
Connector	2						
Coupling	1½						
Strap	2						
Fasteners	3						
#12 THHN wire	38'						
Wire nuts	2						
Misc., tapes, wire lube, etc.							
Material							
Sales Tax							
Material Subtotal							
Man-hours							
Labor Hourly Rate							
Labor Subtotal							
Material and Labor Subtotal							
Allow 30% for Overhead and Profit							

FIGURE 7.20 — UNIT PRICING FORMS FOR COMMERCIAL CONSTRUCTION (cont.)

THREE-WAY SWITCH

Description	Quantity	Material			Labor		
		Unit Price	Per	Extension	Hours	Per	Extension
3-way switch	1						
Plate	1						
1-gg, 1/2" raise ring	1						
4" square box	1						
1/2" EMT	20'						
Connector	2						
Coupling	1½						
Strap	2½						
Fasteners	3						
#12 THHN wire	100'						
Wire nuts	4						
Tapes, etc.							
Material							
Sales Tax							
Material Subtotal							
Man-hours							
Labor Hourly Rate							
Labor Subtotal							
Material and Labor Subtotal							
Allow 30% for Overhead and Profit							
Total Unit Price							

FIGURE 7.20 — UNIT PRICING FORMS FOR COMMERCIAL CONSTRUCTION *(cont.)*

2 × 4 LAY-IN FIXTURE

Description	Quantity	Material			Labor		
		Unit Price	Per	Extension	Hours	Per	Extension
2 × 4 fixture, pre-lamped, pre-whipped	1						
Fixture clips	2						
4" square box	1						
Blank cover	1						
Fasteners	2						
1/2" EMT	12'						
Connector	2						
Coupling	1						
Strap	1 1/2						
#12 THHN wire	40'						
Wire nuts	3						
Circuit breaker	1/6						
Tapes, etc.							
Material							
Sales Tax							
Material Subtotal							
Man-hours							
Labor Hourly Rate							
Labor Subtotal							
Material and Labor Subtotal							
Allow 30% for Overhead and Profit							

7-42

FIGURE 7.20 – UNIT PRICING FORMS FOR COMMERCIAL CONSTRUCTION (cont.)

PHONE STUB

Description	Quantity	Material Unit Price	Material Per	Material Extension	Labor Hours	Labor Per	Labor Extension
1/2" EMT	10'						
Connector	1						
Strap	1						
4" square box	1						
1-gg, 1/2" raise ring	1						
Fasteners	3						
Tapes, etc.							
Material							
Sales Tax							
Material Subtotal							
Man-hours							
Labor Hourly Rate							
Labor Subtotal							
Material and Labor Subtotal							
Allow 30% for Overhead and Profit							
Total Unit Price							

FIGURE 7.20 – UNIT PRICING FORMS FOR COMMERCIAL CONSTRUCTION

3-PHASE ROOFTOP UNIT

Description	Quantity	Material Unit Price	Material Per	Material Extension	Labor Hours	Labor Per	Labor Extension
60 amp, 3-pole, non-fused disconnect	1						
3/4" chase nipple	1						
Locknut	1						
3/4" Sealtite	4'						
Connectors – Straight	2						
Pipe coupling	1						
Fasteners	3						
Terminations	4						
Wire nuts	4						
3/4" EMT	50'						
Connector	1						
Conn-compression	1						
Coupling	5						
Strap	6						
#6 THHN wire	186'						
#10 THHN wire	62'						
3-pole circuit breaker	1						
Tapes, etc.							

Material

Sales Tax

Material Subtotal

Man-hours

Labor Hourly Rate

Labor Subtotal

Material and Labor Subtotal

Allow 30% for Overhead and Profit

Total Unit Price

FIGURE 7.20 — UNIT PRICING FORMS FOR COMMERCIAL CONSTRUCTION (cont.)

200 AMP, 3-PHASE OVERHEAD SERVICE

Description	Quantity	Material			Labor		
		Unit Price	Per	Extension	Hours	Per	Extension
2½" GRC	10'						
Head	1						
Back plates	2						
Straps	2						
Hub	1						
Nipple	1						
Locknuts	4						
Bushings	2						
Meter (furnished by utility)	1						
MCB panel-42 space	1						
#250 aluminum THW wire	60'						
#2 aluminum THW wire	20'						
1/2" EMT	30'						
Connector	3						
Coupling	3						
Strap	4						
Ground rod	1						
Ground rod clamp	1						
Water pipe clamp	1						
#4 bare copper wire	45'						

Fasteners	18					
Cut and patch wall	1					
Tapes, etc.						
Material						
Sales Tax						
Material Subtotal						
Man-hours						
Labor Hourly Rate						
Labor Subtotal						
Material and Labor Subtotal						
Allow 50% for Overhead and Profit						
Total Unit Price						

FIGURE 7.20 – UNIT PRICING FORMS FOR COMMERCIAL CONSTRUCTION *(cont.)*

EMERGENCY LIGHT

Description	Quantity	Material			Labor		
		Unit Price	Per	Extension	Hours	Per	Extension
2-head emergency light	1						
4" square box	1						
1-gg, 1/2" raise ring	1						
Fasteners	3						
1/2" EMT	16'						
Connector	2						
Coupling	1½						
Strap	2						
#12 THHN wire	39'						
Wire nuts	2						
Tapes, etc.	1/10						
Material							
Sales Tax							
Material Subtotal							
Man-hours							
Labor Hourly Rate							
Labor Subtotal							
Material and Labor Subtotal							
Allow 30% for Overhead and Profit							
Total Unit Price							

FIGURE 7.20 — UNIT PRICING FORMS FOR COMMERCIAL CONSTRUCTION *(cont.)*

BATHROOM LIGHT

Description	Quantity	Material				Labor		
		Unit Price	Per	Extension		Hours	Per	Extension
Wall-mounted incandescent fixture with lamp	1							
4" square box	1							
Round ring	1							
1/2" EMT	15'							
Connector	2							
Coupling	1½							
Strap	2							
Fasteners	3							
#12 THHN Wire	36'							
Wire nuts	3							
Circuit breaker	1/6							
Tapes, etc.								
Material								
Sales Tax								
Material Subtotal								
Man-hours								
Labor Hourly Rate								
Labor Subtotal								
Material and Labor Subtotal								
Allow 30% for Overhead and Profit								
Total Unit Price								

7-49

NOTES

CHAPTER 8
Technical Reference

Disclaimer: Data tables in this chapter represent the author's best judgment and care for the information published. Instructions for these data tables, when given, should be carefully studied before using them. The numerical results for any data table is affected by numerous specific project factors such as design standards, site conditions, labor productivity, material waste, etc. Please also refer to the governing electrical codes for your specific jobs. If anything in this book conflicts with your code, the code should always govern. Neither the author, DeWALT, nor the publisher is responsible for any losses or damages with respect to the accuracy, correctness, value and sufficiency of the data, methods, and other information contained herein.

AREAS OF COMMON GEOMETRIC SHAPES

Shape	Formula
	Parallelogram Area = a × b
	Trapezoid Area = c × 1/2 × (a + b)
  Hypotenuse (c) = $\sqrt{a^2 + b^2}$	**Right Triangle** Area = 1/2 × a × b
	Regular Triangle Area = 1/2 × a × b
  b (Diameter) = 2 × a (Radius)	**Circle** Area = 3.1416 × a^2 Circumference = 3.1416 × b = 6.2832 × a

VOLUMES OF COMMON GEOMETRIC SHAPES

Shape	Formula
	Cylinder Volume $= 3.1416 \times \frac{a}{2} \times \frac{a}{2} \times b$ $= 0.7854 \times a^2 \times b$
	Pyramid Volume $= \frac{1}{3} \times a \times b \times c$
	Cone Volume $= \frac{1}{3} \times 3.1416 \times b \times b \times c$ $= 1.0472 \times b^2 \times c$ Or Volume $= 0.3518 \times a^2 \times c$
	Sphere Volume $= \frac{1}{6} \times 3.1416 \times a \times a \times a$ $= 0.5236 \times a^3$

8-3

SQUARE, CUBE, SQUARE ROOT, AND CUBIC ROOT FOR NUMBERS FROM 1 TO 100

Number	Square	Cube	Square Root	Cubic Root
1	1	1	1.0000	1.0000
2	4	8	1.4142	1.2599
3	9	27	1.7321	1.4422
4	16	64	2.0000	1.5874
5	25	125	2.2361	1.7100
6	36	216	2.4495	1.8171
7	49	343	2.6458	1.9129
8	64	512	2.8284	2.0000
9	81	729	3.0000	2.0801
10	100	1000	3.1623	2.1544
11	121	1331	3.3166	2.2240
12	144	1728	3.4641	2.2894
13	169	2197	3.6056	2.3513
14	196	2744	3.7417	2.4101
15	225	3375	3.8730	2.4662
16	256	4096	4.0000	2.5198
17	289	4913	4.1231	2.5713
18	324	5832	4.2426	2.6207
19	361	6859	4.3589	2.6684
20	400	8000	4.4721	2.7144
21	441	9261	4.5826	2.7589
22	484	10648	4.6904	2.8020
23	529	12167	4.7958	2.8439
24	576	13824	4.8990	2.8845
25	625	15625	5.0000	2.9240
26	676	17576	5.0990	2.9625
27	729	19683	5.1962	3.0000
28	784	21952	5.2915	3.0366
29	841	24389	5.3852	3.0723
30	900	27000	5.4772	3.1072
31	961	29791	5.5678	3.1414
32	1024	32768	5.6569	3.1748
33	1089	35937	5.7446	3.2075

Number	Square	Cube	Square Root	Cubic Root
34	1156	39304	5.8310	3.2396
35	1225	42875	5.9161	3.2711
36	1296	46656	6.0000	3.3019
37	1369	50653	6.0828	3.3322
38	1444	54872	6.1644	3.3620
39	1521	59319	6.2450	3.3912
40	1600	64000	6.3246	3.4200
41	1681	68921	6.4031	3.4482
42	1764	74088	6.4807	3.4760
43	1849	79507	6.5574	3.5034
44	1936	85184	6.6332	3.5303
45	2025	91125	6.7082	3.5569
46	2116	97336	6.7823	3.5830
47	2209	103823	6.8557	3.6088
48	2304	110592	6.9282	3.6342
49	2401	117649	7.0000	3.6593
50	2500	125000	7.0711	3.6840
51	2601	132651	7.1414	3.7084
52	2704	140608	7.2111	3.7325
53	2809	148877	7.2801	3.7563
54	2916	157464	7.3485	3.7798
55	3025	166375	7.4162	3.8030
56	3136	175616	7.4833	3.8259
57	3249	185193	7.5498	3.8485
58	3364	195112	7.6158	3.8709
59	3481	205379	7.6811	3.8930
60	3600	216000	7.7460	3.9149
61	3721	226981	7.8102	3.9365
62	3844	238328	7.8740	3.9579
63	3969	250047	7.9373	3.9791
64	4096	262144	8.0000	4.0000
65	4225	274625	8.0623	4.0207
66	4356	287496	8.1240	4.0412

SQUARE, CUBE, SQUARE ROOT, AND CUBIC ROOT
FOR NUMBERS FROM 1 TO 100 *(cont.)*

Number	Square	Cube	Square Root	Cubic Root
67	4489	300763	8.1854	4.0615
68	4624	314432	8.2462	4.0817
69	4761	328509	8.3066	4.1016
70	4900	343000	8.3666	4.1213
71	5041	357911	8.4261	4.1408
72	5184	373248	8.4853	4.1602
73	5329	389017	8.5440	4.1793
74	5476	405224	8.6023	4.1983
75	5625	421875	8.6603	4.2172
76	5776	438976	8.7178	4.2358
77	5929	456533	8.7750	4.2543
78	6084	474552	8.8318	4.2727
79	6241	493039	8.8882	4.2908
80	6400	512000	8.9443	4.3089
81	6561	531441	9.0000	4.3267
82	6724	551368	9.0554	4.3445
83	6889	571787	9.1104	4.3621
84	7056	592704	9.1652	4.3795
85	7225	614125	9.2195	4.3968
86	7396	636056	9.2736	4.4140
87	7569	658503	9.3274	4.4310
88	7744	681472	9.3808	4.4480
89	7921	704969	9.4340	4.4647
90	8100	729000	9.4868	4.4814
91	8281	753571	9.5394	4.4979
92	8464	778688	9.5917	4.5144
93	8649	804357	9.6437	4.5307
94	8836	830584	9.6954	4.5468
95	9025	857375	9.7468	4.5629
96	9216	884736	9.7980	4.5789
97	9409	912673	9.8489	4.5947
98	9604	941192	9.8995	4.6104
99	9801	970299	9.9499	4.6261
100	10000	1000000	10.0000	4.6416

CIRCLE CIRCUMFERENCE AND AREA
(DIAMETERS FROM 1 TO 100)

Diameter	Radius	Circumference	Area
		(based on the same type of units)	
1	0.5	3.1416	0.7854
2	1.0	6.2832	3.1416
3	1.5	9.4248	7.0686
4	2.0	12.5664	12.5664
5	2.5	15.7080	19.6350
6	3.0	18.8496	28.2744
7	3.5	21.9912	38.4846
8	4.0	25.1328	50.2656
9	4.5	28.2744	63.6174
10	5.0	31.4160	78.5400
11	5.5	34.5576	95.0334
12	6.0	37.6992	113.0976
13	6.5	40.8408	132.7326
14	7.0	43.9824	153.9384
15	7.5	47.1240	176.7150
16	8.0	50.2656	201.0624
17	8.5	53.4072	226.9806
18	9.0	56.5488	254.4696
19	9.5	59.6904	283.5294
20	10.0	62.8320	314.1600
21	10.5	65.9736	346.3614
22	11.0	69.1152	380.1336
23	11.5	72.2568	415.4766
24	12.0	75.3984	452.3904
25	12.5	78.5400	490.8750
26	13.0	81.6816	530.9304
27	13.5	84.8232	572.5566
28	14.0	87.9648	615.7536
29	14.5	91.1064	660.5214
30	15.0	94.2480	706.8600
31	15.5	97.3896	754.7694
32	16.0	100.5312	804.2496
33	16.5	103.6728	855.3006

Diameter	Radius	Circumference	Area
(based on the same type of units)			
34	17.0	106.8144	907.9224
35	17.5	109.9560	962.1150
36	18.0	113.0976	1017.8784
37	18.5	116.2392	1075.2126
38	19.0	119.3808	1134.1176
39	19.5	122.5224	1194.5934
40	20.0	125.6640	1256.6400
41	20.5	128.8056	1320.2574
42	21.0	131.9472	1385.4456
43	21.5	135.0888	1452.2046
44	22.0	138.2304	1520.5344
45	22.5	141.3720	1590.4350
46	23.0	144.5136	1661.9064
47	23.5	147.6552	1734.9486
48	24.0	150.7968	1809.5616
49	24.5	153.9384	1885.7454
50	25.0	157.0800	1963.5000
51	25.5	160.2216	2042.8254
52	26.0	163.3632	2123.7216
53	26.5	166.5048	2206.1886
54	27.0	169.6464	2290.2264
55	27.5	172.7880	2375.8350
56	28.0	175.9296	2463.0144
57	28.5	179.0712	2551.7646
58	29.0	182.2128	2642.0856
59	29.5	185.3544	2733.9774
60	30.0	188.4960	2827.4400
61	30.5	191.6376	2922.4734
62	31.0	194.7792	3019.0776
63	31.5	197.9208	3117.2526
64	32.0	201.0624	3216.9984
65	32.5	204.2040	3318.3150
66	33.0	207.3456	3421.2024

CIRCLE CIRCUMFERENCE AND AREA
(DIAMETERS FROM 1 TO 100) (cont.)

Diameter	Radius	Circumference	Area
		(based on the same type of units)	
67	33.5	210.4872	3525.6606
68	34.0	213.6288	3631.6896
69	34.5	216.7704	3739.2894
70	35.0	219.9120	3848.4600
71	35.5	223.0536	3959.2014
72	36.0	226.1952	4071.5136
73	36.5	229.3368	4185.3966
74	37.0	232.4784	4300.8504
75	37.5	235.6200	4417.8750
76	38.0	238.7616	4536.4704
77	38.5	241.9032	4656.6366
78	39.0	245.0448	4778.3736
79	39.5	248.1864	4901.6814
80	40.0	251.3280	5026.5600
81	40.5	254.4696	5153.0094
82	41.0	257.6112	5281.0296
83	41.5	260.7528	5410.6206
84	42.0	263.8944	5541.7824
85	42.5	267.0360	5674.5150
86	43.0	270.1776	5808.8184
87	43.5	273.3192	5944.6926
88	44.0	276.4608	6082.1376
89	44.5	279.6024	6221.1534
90	45.0	282.7440	6361.7400
91	45.5	285.8856	6503.8974
92	46.0	289.0272	6647.6256
93	46.5	292.1688	6792.9246
94	47.0	295.3104	6939.7944
95	47.5	298.4520	7088.2350
96	48.0	301.5936	7238.2464
97	48.5	304.7352	7389.8286
98	49.0	307.8768	7542.9816
99	49.5	311.0184	7697.7054
100	50.0	314.1600	7854.0000

TRIGONOMETRIC FUNCTIONS

Degree	Sin	Cos	Tan	Cot	Sec	Csc
0	0.0000	1.0000	0.0000	Infinity	1.0000	Infinity
1/2	0.0087	1.0000	0.0087	114.5887	1.0000	114.5930
1	0.0175	0.9998	0.0175	57.2900	1.0002	57.2987
1 1/2	0.0262	0.9997	0.0262	38.1885	1.0003	38.2016
2	0.0349	0.9994	0.0349	28.6363	1.0006	28.6537
2 1/2	0.0436	0.9990	0.0437	22.9038	1.0010	22.9256
3	0.0523	0.9986	0.0524	19.0811	1.0014	19.1073
3 1/2	0.0610	0.9981	0.0612	16.3499	1.0019	16.3804
4	0.0698	0.9976	0.0699	14.3007	1.0024	14.3356
4 1/2	0.0785	0.9969	0.0787	12.7062	1.0031	12.7455
5	0.0872	0.9962	0.0875	11.4301	1.0038	11.4737
5 1/2	0.0958	0.9954	0.0963	10.3854	1.0046	10.4334
6	0.1045	0.9945	0.1051	9.5144	1.0055	9.5668
6 1/2	0.1132	0.9936	0.1139	8.7769	1.0065	8.8337
7	0.1219	0.9925	0.1228	8.1443	1.0075	8.2055
7 1/2	0.1305	0.9914	0.1317	7.5958	1.0086	7.6613
8	0.1392	0.9903	0.1405	7.1154	1.0098	7.1853
8 1/2	0.1478	0.9890	0.1495	6.6912	1.0111	6.7655
9	0.1564	0.9877	0.1584	6.3138	1.0125	6.3925
9 1/2	0.1650	0.9863	0.1673	5.9758	1.0139	6.0589
10	0.1736	0.9848	0.1763	5.6713	1.0154	5.7588
10 1/2	0.1822	0.9833	0.1853	5.3955	1.0170	5.4874
11	0.1908	0.9816	0.1944	5.1446	1.0187	5.2408
11 1/2	0.1994	0.9799	0.2035	4.9152	1.0205	5.0159
12	0.2079	0.9781	0.2126	4.7046	1.0223	4.8097
12 1/2	0.2164	0.9763	0.2217	4.5107	1.0243	4.6202
13	0.2250	0.9744	0.2309	4.3315	1.0263	4.4454
13 1/2	0.2334	0.9724	0.2401	4.1653	1.0284	4.2837
14	0.2419	0.9703	0.2493	4.0108	1.0306	4.1336
14 1/2	0.2504	0.9681	0.2586	3.8667	1.0329	3.9939
15	0.2588	0.9659	0.2679	3.7321	1.0353	3.8637
15 1/2	0.2672	0.9636	0.2773	3.6059	1.0377	3.7420

TRIGONOMETRIC FUNCTIONS (cont.)

Degree	Sin	Cos	Tan	Cot	Sec	Csc
16	0.2756	0.9613	0.2867	3.4874	1.0403	3.6280
16½	0.2840	0.9588	0.2962	3.3759	1.0429	3.5209
17	0.2924	0.9563	0.3057	3.2709	1.0457	3.4203
17½	0.3007	0.9537	0.3153	3.1716	1.0485	3.3255
18	0.3090	0.9511	0.3249	3.0777	1.0515	3.2361
18½	0.3173	0.9483	0.3346	2.9887	1.0545	3.1515
19	0.3256	0.9455	0.3443	2.9042	1.0576	3.0716
19½	0.3338	0.9426	0.3541	2.8239	1.0608	2.9957
20	0.3420	0.9397	0.3640	2.7475	1.0642	2.9238
20½	0.3502	0.9367	0.3739	2.6746	1.0676	2.8555
21	0.3584	0.9336	0.3839	2.6051	1.0711	2.7904
21½	0.3665	0.9304	0.3939	2.5386	1.0748	2.7285
22	0.3746	0.9272	0.4040	2.4751	1.0785	2.6695
22½	0.3827	0.9239	0.4142	2.4142	1.0824	2.6131
23	0.3907	0.9205	0.4245	2.3559	1.0864	2.5593
23½	0.3987	0.9171	0.4348	2.2998	1.0904	2.5078
24	0.4067	0.9135	0.4452	2.2460	1.0946	2.4586
24½	0.4147	0.9100	0.4557	2.1943	1.0989	2.4114
25	0.4226	0.9063	0.4663	2.1445	1.1034	2.3662
25½	0.4305	0.9026	0.4770	2.0965	1.1079	2.3228
26	0.4384	0.8988	0.4877	2.0503	1.1126	2.2812
26½	0.4462	0.8949	0.4986	2.0057	1.1174	2.2412
27	0.4540	0.8910	0.5095	1.9626	1.1223	2.2027
27½	0.4617	0.8870	0.5206	1.9210	1.1274	2.1657
28	0.4695	0.8829	0.5317	1.8807	1.1326	2.1301
28½	0.4772	0.8788	0.5430	1.8418	1.1379	2.0957
29	0.4848	0.8746	0.5543	1.8040	1.1434	2.0627
29½	0.4924	0.8704	0.5658	1.7675	1.1490	2.0308
30	0.5000	0.8660	0.5774	1.7321	1.1547	2.0000
30½	0.5075	0.8616	0.5890	1.6977	1.1606	1.9703
31	0.5150	0.8572	0.6009	1.6643	1.1666	1.9416
31½	0.5225	0.8526	0.6128	1.6319	1.1728	1.9139

TRIGONOMETRIC FUNCTIONS (cont.)

Degree	Sin	Cos	Tan	Cot	Sec	Csc
32	0.5299	0.8480	0.6249	1.6003	1.1792	1.8871
32½	0.5373	0.8434	0.6371	1.5697	1.1857	1.8612
33	0.5446	0.8387	0.6494	1.5399	1.1924	1.8361
33½	0.5519	0.8339	0.6619	1.5108	1.1992	1.8118
34	0.5592	0.8290	0.6745	1.4826	1.2062	1.7883
34½	0.5664	0.8241	0.6873	1.4550	1.2134	1.7655
35	0.5736	0.8192	0.7002	1.4281	1.2208	1.7434
35½	0.5807	0.8141	0.7133	1.4019	1.2283	1.7221
36	0.5878	0.8090	0.7265	1.3764	1.2361	1.7013
36½	0.5948	0.8039	0.7400	1.3514	1.2440	1.6812
37	0.6018	0.7986	0.7536	1.3270	1.2521	1.6616
37½	0.6088	0.7934	0.7673	1.3032	1.2605	1.6427
38	0.6157	0.7880	0.7813	1.2799	1.2690	1.6243
38½	0.6225	0.7826	0.7954	1.2572	1.2778	1.6064
39	0.6293	0.7771	0.8098	1.2349	1.2868	1.5890
39½	0.6361	0.7716	0.8243	1.2131	1.2960	1.5721
40	0.6428	0.7660	0.8391	1.1918	1.3054	1.5557
40½	0.6494	0.7604	0.8541	1.1708	1.3151	1.5398
41	0.6561	0.7547	0.8693	1.1504	1.3250	1.5243
41½	0.6626	0.7490	0.8847	1.1303	1.3352	1.5092
42	0.6691	0.7431	0.9004	1.1106	1.3456	1.4945
42½	0.6756	0.7373	0.9163	1.0913	1.3563	1.4802
43	0.6820	0.7314	0.9325	1.0724	1.3673	1.4663
43½	0.6884	0.7254	0.9490	1.0538	1.3786	1.4527
44	0.6947	0.7193	0.9657	1.0355	1.3902	1.4396
44½	0.7009	0.7133	0.9827	1.0176	1.4020	1.4267
45	0.7071	0.7071	1.0000	1.0000	1.4142	1.4142
45½	0.7133	0.7009	1.0176	0.9827	1.4267	1.4020
46	0.7193	0.6947	1.0355	0.9657	1.4396	1.3902
46½	0.7254	0.6884	1.0538	0.9490	1.4527	1.3786
47	0.7314	0.6820	1.0724	0.9325	1.4663	1.3673
47½	0.7373	0.6756	1.0913	0.9163	1.4802	1.3563

TRIGONOMETRIC FUNCTIONS (cont.)

Degree	Sin	Cos	Tan	Cot	Sec	Csc
48	0.7431	0.6691	1.1106	0.9004	1.4945	1.3456
48¹/₂	0.7490	0.6626	1.1303	0.8847	1.5092	1.3352
49	0.7547	0.6561	1.1504	0.8693	1.5243	1.3250
49¹/₂	0.7604	0.6494	1.1708	0.8541	1.5398	1.3151
50	0.7660	0.6428	1.1918	0.8391	1.5557	1.3054
50¹/₂	0.7716	0.6361	1.2131	0.8243	1.5721	1.2960
51	0.7771	0.6293	1.2349	0.8098	1.5890	1.2868
51¹/₂	0.7826	0.6225	1.2572	0.7954	1.6064	1.2778
52	0.7880	0.6157	1.2799	0.7813	1.6243	1.2690
52¹/₂	0.7934	0.6088	1.3032	0.7673	1.6427	1.2605
53	0.7986	0.6018	1.3270	0.7536	1.6616	1.2521
53¹/₂	0.8039	0.5948	1.3514	0.7400	1.6812	1.2440
54	0.8090	0.5878	1.3764	0.7265	1.7013	1.2361
54¹/₂	0.8141	0.5807	1.4019	0.7133	1.7221	1.2283
55	0.8192	0.5736	1.4281	0.7002	1.7434	1.2208
55¹/₂	0.8241	0.5664	1.4550	0.6873	1.7655	1.2134
56	0.8290	0.5592	1.4826	0.6745	1.7883	1.2062
56¹/₂	0.8339	0.5519	1.5108	0.6619	1.8118	1.1992
57	0.8387	0.5446	1.5399	0.6494	1.8361	1.1924
57¹/₂	0.8434	0.5373	1.5697	0.6371	1.8612	1.1857
58	0.8480	0.5299	1.6003	0.6249	1.8871	1.1792
58¹/₂	0.8526	0.5225	1.6319	0.6128	1.9139	1.1728
59	0.8572	0.5150	1.6643	0.6009	1.9416	1.1666
59¹/₂	0.8616	0.5075	1.6977	0.5890	1.9703	1.1606
60	0.8660	0.5000	1.7321	0.5774	2.0000	1.1547
60¹/₂	0.8704	0.4924	1.7675	0.5658	2.0308	1.1490
61	0.8746	0.4848	1.8040	0.5543	2.0627	1.1434
61¹/₂	0.8788	0.4772	1.8418	0.5430	2.0957	1.1379
62	0.8829	0.4695	1.8807	0.5317	2.1301	1.1326
62¹/₂	0.8870	0.4617	1.9210	0.5206	2.1657	1.1274
63	0.8910	0.4540	1.9626	0.5095	2.2027	1.1223
63¹/₂	0.8949	0.4462	2.0057	0.4986	2.2412	1.1174

TRIGONOMETRIC FUNCTIONS (cont.)

Degree	Sin	Cos	Tan	Cot	Sec	Csc
64	0.8988	0.4384	2.0503	0.4877	2.2812	1.1126
64¹/₂	0.9026	0.4305	2.0965	0.4770	2.3228	1.1079
65	0.9063	0.4226	2.1445	0.4663	2.3662	1.1034
65¹/₂	0.9100	0.4147	2.1943	0.4557	2.4114	1.0989
66	0.9135	0.4067	2.2460	0.4452	2.4586	1.0946
66¹/₂	0.9171	0.3987	2.2998	0.4348	2.5078	1.0904
67	0.9205	0.3907	2.3559	0.4245	2.5593	1.0864
67¹/₂	0.9239	0.3827	2.4142	0.4142	2.6131	1.0824
68	0.9272	0.3746	2.4751	0.4040	2.6695	1.0785
68¹/₂	0.9304	0.3665	2.5386	0.3939	2.7285	1.0748
69	0.9336	0.3584	2.6051	0.3839	2.7904	1.0711
69¹/₂	0.9367	0.3502	2.6746	0.37388	2.8555	1.0676
70	0.9397	0.3420	2.7475	0.36397	2.9238	1.0642
70¹/₂	0.9426	0.3338	2.8239	0.35412	2.9957	1.0608
71	0.9455	0.3256	2.9042	0.34433	3.0716	1.0576
71¹/₂	0.9483	0.3173	2.9887	0.33460	3.1515	1.0545
72	0.9511	0.3090	3.0777	0.32492	3.2361	1.0515
72¹/₂	0.9537	0.3007	3.1716	0.31530	3.3255	1.0485
73	0.9563	0.2924	3.2709	0.30573	3.4203	1.0457
73¹/₂	0.9588	0.2840	3.3759	0.29621	3.5209	1.0429
74	0.9613	0.2756	3.4874	0.28675	3.6280	1.0403
74¹/₂	0.9636	0.2672	3.6059	0.27732	3.7420	1.0377
75	0.9659	0.2588	3.7321	0.26795	3.8637	1.0353
75¹/₂	0.9681	0.2504	3.8667	0.25862	3.9939	1.0329
76	0.9703	0.2419	4.0108	0.24933	4.1336	1.0306
76¹/₂	0.9724	0.2334	4.1653	0.24008	4.2837	1.0284
77	0.9744	0.2250	4.3315	0.23087	4.4454	1.0263
77¹/₂	0.9763	0.2164	4.5107	0.22169	4.6202	1.0243
78	0.9781	0.2079	4.7046	0.21256	4.8097	1.0223
78¹/₂	0.9799	0.1994	4.9152	0.20345	5.0159	1.0205
79	0.9816	0.1908	5.1446	0.19438	5.2408	1.0187
79¹/₂	0.9833	0.1822	5.3955	0.18534	5.4874	1.0170

TRIGONOMETRIC FUNCTIONS *(cont.)*

Degree	Sin	Cos	Tan	Cot	Sec	Csc
80	0.9848	0.1736	5.6713	0.17633	5.7588	1.0154
80½	0.9863	0.1650	5.9758	0.16734	6.0589	1.0139
81	0.9877	0.1564	6.3138	0.15838	6.3925	1.0125
81½	0.9890	0.1478	6.6912	0.14945	6.7655	1.0111
82	0.9903	0.1392	7.1154	0.14054	7.1853	1.0098
82½	0.9914	0.1305	7.5958	0.13165	7.6613	1.0086
83	0.9925	0.1219	8.1443	0.12278	8.2055	1.0075
83½	0.9936	0.1132	8.7769	0.11394	8.8337	1.0065
84	0.9945	0.1045	9.5144	0.10510	9.5668	1.0055
84½	0.9954	0.0958	10.3854	0.09629	10.4334	1.0046
85	0.9962	0.0872	11.4301	0.08749	11.4737	1.0038
85 ½	0.9969	0.0785	12.7062	0.07870	12.7455	1.0031
86	0.9976	0.0698	14.3007	0.06993	14.3356	1.0024
86 ½	0.9981	0.0610	16.3499	0.06116	16.3804	1.0019
87	0.9986	0.0523	19.0811	0.05241	19.1073	1.0014
87 ½	0.9990	0.0436	22.9038	0.04366	22.9256	1.0010
88	0.9994	0.0349	28.6363	0.03492	28.6537	1.0006
88 ½	0.9997	0.0262	38.1885	0.02619	38.2016	1.0003
89	0.9998	0.0175	57.2900	0.01746	57.2987	1.0002
89 ½	1.0000	0.0087	114.5887	0.00873	114.5930	1.0000
90	1.0000	0.0000	Infinity	0.00000	Infinity	1.0000

MASTER-FORMAT 1995 EDITION

Series 0 - Bidding and Contracting Requirements
Division 01 General Requirements
Division 02 Site Work
Division 03 Concrete
Division 04 Masonry
Division 05 Metals
Division 06 Wood, Plastics
Division 07 Thermal and Moisture Protection
Division 08 Doors and Windows
Division 09 Finishes
Division 10 Specialties
Division 11 Equipment
Division 12 Furnishings
Division 13 Special Construction
Division 14 Conveying Equipment
Division 15 Mechanical
Division 16 Electrical

MASTER-FORMAT 1995 EDITION ELECTRICAL TITLES

Main Titles
Division 16 Electrical
    16050 Basic Electrical Materials and Methods
    16100 Wiring Methods
    16200 Electrical Power
    16300 Transmission and Distribution
    16400 Low-Voltage Distribution
    16500 Lighting
    16700 Communications
    16800 Sound and Video

Code	Short Description	Extended Description
16050	**Basic Electrical Materials and Methods**	
16060	Grounding and Bonding	—
16070	Hangers and Support	Seismic Controls
16075	Electrical Identification	—
16080	Electrical Testing	Maintenance Testing
16090	Restoration and Repair	—
16100	**Wiring Method**	
16120	Conductors and Cable	—
16130	Raceway and Boxes	Cabinets Conduit and Tubing
		Cutout Boxes
		Enclosures
		Indoor Service Poles
		Junction Boxes
		Multi-Outlet Assemblies
		Outlet Boxes
		Pull Boxes
		Surface Raceway
		Wireway and Auxiliary Gutters
16140	Wiring Device	Floor Boxes
		Receptacles
		Remote Control Switching Devices
		Wall Plates
		Wall Switches and Dimmers
16150	Wiring Connection	—
16200	**Electrical Power**	
16210	Electrical Utility Service	—

Code	Short Description	Extended Description
16220	Motors and Generators	—
16230	Generator Assemblies	Engine Generators
		Frequency Changers
		Motor Generators
		Rotary Converters
		Rotary Uninterruptible Power Units
16240	Battery Equipment	Batteries
		Battery Racks
		Battery Units
		Central Battery Equipment
16260	Static Power Converter	Battery Chargers
		DC Drive Controllers
		Slip Controllers
		Static Frequency Converters
		Static Uninterruptible Power Supplies
		Variable Frequency Controllers
16270	Transformer	Distribution Transformers
		Network Transformers
		Pad-Mounted Transformers
		Power Transformers
		Substation Transformers
16280	Power Filters and Conditioner	Capacitors
		Chokes and Inductors
		EMI Filters
		Harmonic Filters

Code	Short Description	Extended Description
16280	Power Filters and Conditioner *(cont.)*	Power Factor Controllers
		RFI Filters
		Surge Suppressors
		Voltage Regulators
16290	Power Measurement and Control	—
16300	**Transmission and Distribution**	
16310	Transmission and Distribution Accessories	Arresters
		Cutouts
		Insulators
		Line Materials
		Supports
16320	High-Voltage Switching and Protection	High-Voltage Circuit Breakers
		High-Voltage Cutouts
		High-Voltage Fuses
		High-Voltage Lightning Arresters
		High-Voltage Reclosers
		High-Voltage Surge Arresters
16330	Medium-Voltage Switching and Protection Assemblies	Medium-Voltage Circuit Protection Devices
		Medium-Voltage Cutouts
		Medium-Voltage Fuses
		Medium-Voltage Lightning Arresters
		Medium-Voltage Reclosers
		Medium-Voltage Surge Arresters

Code	Short Description	Extended Description
16340	Medium-Voltage Switching and Protection Assemblies	Medium-Voltage Circuit Breaker Switchgear
		Medium-Voltage Enclosed Bu
		Medium-Voltage Enclosed Fuse Cutouts
		Medium-Voltage Enclosed Fuses
		Medium-Voltage Fusible Interrupter Switchgear
		Medium-Voltage Motor Controllers
		Medium-Voltage Vacuum Interrupter Switchgear
16360	Unit Substation	—
16400	**Low-Voltage Distribution**	
16410	Enclosed Switches and Circuit Breaker	—
16420	Enclosed Controller	—
16430	Low-Voltage Switchgear	—
16440	Switchboards, Panelboards and Control Centers	—
16450	Enclosed Bus Assemblies	—
16460	Low-Voltage Transformer	—
16470	Power Distribution Units	—
16490	Low-Voltage Distribution Components and Accessories	—
16500	**Lighting**	
16510	Interior Luminaries	—

Code	Short Description	Extended Description
16520	Exterior Luminaries	Area Lighting
		Aviation Lighting
		Flood Lighting
		Navigation Lighting
		Parking Lighting
		Roadway Lighting
		Site Lighting
		Sports Lighting
		Walkway Lighting
16530	Emergency Lighting	—
16540	Classified Location Lighting	—
16550	Special-Purpose Lighting	Detention Lighting
		Display Lighting
		Medical Lighting
		Outline Lighting
		Security Lighting
		Theatrical Lighting
		Underwater Lighting
16560	Signal Lighting	Hazard Warning Lighting
		Obstruction Lighting
16570	Dimming Control	—
16580	Lighting Accessories	—
16590	Lighting Restoration and Repair	—
16700	**Communication**	
16710	Communications Circuit	—

MASTER-FORMAT 1995 EDITION
ELECTRICAL TITLES (cont.)

Code	Short Description	Extended Description
16720	Telephone/Intercom Equipment	—
16740	Communication and Data Processing	—
16770	Cable Transmission and Reception	—
16780	Broadcast Transmission and Reception Equipment	—
16790	Microwave Transmission and Reception Equipment	—
16800	**Sound and Video**	
16810	Sound and Video Circuit	—
16820	Sound Reinforcement	—
16830	Broadcast Studio Audio Equipment	—
16840	Broadcast Studio Video Equipment	—
16850	Television Equipment	—
16880	Multimedia Equipment	—

MASTER-FORMAT 2004 EDITION

PROCUREMENT AND CONTRACTING REQUIREMENTS GROUP

Division 00 Procurement and Contracting Requirements

SPECIFICATIONS GROUP

General Requirements Subgroup

Division 01 General Requirements

Facility Construction Subgroup

Division 02 Existing Conditions

Division 03 Concrete

Division 04 Masonry

Division 05 Metals

Division 06 Wood, Plastics, and Composites

Division 07 Thermal and Moisture Protection

Division 08 Openings

Division 09 Finishes

Division 10 Specialties

Division 11 Equipment

Division 12 Furnishings

Division 13 Special Construction

Division 14 Conveying Equipment

Division 15 Reserved

Division 16 Reserved

Division 17 Reserved

MASTER-FORMAT 2004 EDITION *(cont.)*

Facility Construction Subgroup *(cont.)*

Division 18 Reserved

Division 19 Reserved

Facility Services Subgroup

Division 20 Reserved

Division 21 Fire Suppression

Division 22 Plumbing

Division 23 Heating, Ventilating, and Air Conditioning

Division 24 Reserved

Division 25 Integrated Automation

Division 26 Electrical

Division 27 Communications

Division 28 Electronic Safety and Security

Division 29 Reserved

Site and Infrastructure Subgroup

Division 30 Reserved

Division 31 Earthwork

Division 32 Exterior Improvements

Division 33 Utilities

Division 34 Transportation

Division 35 Waterway and Marine Construction

Division 36 Reserved

Division 37 Reserved

Site and Infrastructure Subgroup *(cont.)*

Division 38 Reserved

Division 39 Reserved

Process Equipment Subgroup

Division 40 Process Integration

Division 41 Material Processing and Handling
Equipment

Division 42 Process Heating, Cooling, and
Drying Equipment

Division 43 Process Gas and Liquid Handling,
Purification, and Storage Equipment

Division 44 Pollution Control Equipment

Division 45 Industry-Specific Manufacturing
Equipment

Division 46 Reserved

Division 47 Reserved

Division 48 Electrical Power Generation

Division 49 Reserved

MASTER-FORMAT 2004 EDITION
ELECTRICAL TITLES

DIVISION 26 ELECTRICAL

26 00 00 Electrical

26 01 00 Operation and Maintenance of Electrical Systems

26 05 00 Common Work Results for Electrical

26 06 00 Schedules for Electrical

26 08 00 Commissioning of Electrical Systems

26 09 00 Instrumentation and Control for Electrical Systems

26 10 00 Medium-Voltage Electrical Distribution

26 11 00 Substations

26 12 00 Medium-Voltage Transformers

26 13 00 Medium-Voltage Switchgear

26 18 00 Medium-Voltage Circuit Protection Devices

26 20 00 Low-Voltage Electrical Distribution

26 21 00 Low-Voltage Overhead Electrical Power Systems

26 22 00 Low-Voltage Transformers

26 23 00 Low-Voltage Switchgear

26 24 00 Switchboards and Panelboards

26 25 00 Enclosed Bus Assemblies

26 26 00 Power Distribution Units

26 27 00 Low-Voltage Distribution Equipment

MASTER-FORMAT 2004 EDITION
ELECTRICAL TITLES *(cont.)*

26 28 00 Low-Voltage Circuit Protective Devices

26 29 00 Low-Voltage Controllers

26 30 00 Facility Electrical Power Generating and Storing Equipment

26 31 00 Photovoltaic Collectors

26 32 00 Packaged Generator Assemblies

26 33 00 Battery Equipment

26 35 00 Power Filters and Conditioners

26 36 00 Transfer Switches

26 40 00 Electrical and Cathodic Protection

26 41 00 Facility Lightning Protection

26 42 00 Cathodic Protection

26 43 00 Transient Voltage Suppression

26 50 00 Lighting

26 51 00 Interior Lighting

26 52 00 Emergency Lighting

26 53 00 Exit Signs

26 54 00 Classified Location Lighting

26 55 00 Special Purpose Lighting

26 56 00 Exterior Lighting

26 60 00 Unassigned

26 70 00 Unassigned

26 80 00 Unassigned

26 90 00 Unassigned

DIVISION 27 COMMUNICATIONS

27 00 00 Communications

27 01 00 Operation and Maintenance of Communications Systems

27 05 00 Common Work Results for Communications

27 06 00 Schedules for Communications

27 08 00 Commissioning of Communications

27 10 00 Structured Cabling

27 11 00 Communications Equipment Room Fittings

27 13 00 Communications Backbone Cabling

27 15 00 Communications Horizontal Cabling

27 16 00 Communications Connecting Cords, Devices, and Adapters

27 20 00 Data Communications

27 21 00 Data Communications Network Equipment

27 22 00 Data Communications Hardware

27 24 00 Data Communications Peripheral Data Equipment

27 25 00 Data Communications Software

27 26 00 Data Communications Programming and Integration Services

27 30 00 Voice Communications

27 31 00 Voice Communications Switching and Routing Equipment

27 32 00 Voice Communications Telephone Sets, Facsimiles, and Modems

27 33 00 Voice Communications Messaging

27 34 00 Call Accounting

27 35 00 Call Management

27 40 00 Audio-Video Communications

27 41 00 Audio-Video Systems

27 42 00 Electronic Digital Systems

27 50 00 Distributed Communications and Monitoring Systems

27 51 00 Distributed Audio-Video Communications Systems

27 52 00 Healthcare Communications and Monitoring Systems

27 53 00 Distributed Systems

27 60 00 Unassigned

27 70 00 Unassigned

27 80 00 Unassigned

27 90 00 Unassigned

DIVISION 28 ELECTRONIC SAFETY AND SECURITY

28 00 00 Electronic Safety and Security

28 01 00 Operation and Maintenance of Electronic Safety and Security

28 05 00 Common Work Results for Electronic Safety and Security

28 06 00 Schedules for Electronic Safety and Security

28 08 00 Commissioning of Electronic Safety and Security

28 10 00 Electronic Access Control and Intrusion Detection

28 13 00 Access Control

28 16 00 Intrusion Detection

28 20 00 Electronic Surveillance

28 23 00 Video Surveillance

28 26 00 Electronic Personal Protection Systems

28 30 00 Electronic Detection and Alarm

28 31 00 Fire Detection and Alarm

28 32 00 Radiation Detection and Alarm

28 33 00 Fuel -Gas Detection and Alarm

28 34 00 Fuel-Oil Detection and Alarm

28 35 00 Refrigerant Detection and Alarm

28 40 00 Electronic Monitoring and Control

28 46 00 Electronic Detention Monitoring and
Control Systems

28 50 00 Unassigned

28 60 00 Unassigned

28 70 00 Unassigned

28 80 00 Unassigned

28 90 00 Unassigned

DIVISION 33 UTILITIES

33 70 00 Electrical Utilities

33 71 00 Electrical Utility Transmission
and Distribution

33 72 00 Utility Substations

33 73 00 Utility Transformers

33 75 00 High-Voltage Switchgear and
Protection Devices

33 77 00 Medium-Voltage Utility Switchgear and
Protection Devices

33 79 00 Site Grounding

33 80 00 Communications Utilities

33 81 00 Communications Structures

33 82 00 Communications Distribution

33 83 00 Wireless Communications Distribution

33 90 00 Unassigned

UNIFORMAT LEVELS AND TITLES

A Substructure
A10   Foundations
A20   Basement Construction
B Shell
B10   Superstructure
B20   Exterior Enclosure
B30   Roofing
C Interiors
C10   Interior Construction
C20   Stairs
C30   Interior Finishes
D Services
D10   Conveying Systems
D20   Plumbing
D30   Heating, Ventilating, and Air Conditioning (HVAC)
D40   Fire Protection Systems
D50   Electrical Systems
E Equipment and Furnishings
E10   Equipment
E20   Furnishings
F Special Construction and Demolition
F10   Special Construction
F20   Selective Demolition
G Building Sitework
G10   Site Preparation
G20   Site Improvements
G30   Site Civil/Mechanical Utilities
G40   Site Electrical Utilities
G90   Other Site Construction
Z General
Z10   General Requirements
Z20   Bidding Requirements, Contract Forms, and
       Conditions Contingencies
Z90   Project Cost Estimate
Project Description
10   Project Description
20   Proposal, Bidding, and Contracting
30   Cost Summary

UNIFORMAT ELECTRICAL TITLES — D 50 ELECTRICAL

D 5010 SERVICE AND DISTRIBUTION

Includes	Excludes
Primary transformers	Outdoor transformers *(see section G 4010, Electrical Distribution)*
Secondary transformers	Emergency power *(see section D 5090, Other Electrical Systems)*
Main switchboard	Branch wiring *(see section D 5020, Lighting and Branch Wiring)*
Interior distribution transformers	—
Branch circuit panels	—
Enclosed circuit breakers	—
Motor control centers	—
Conduit and wiring to circuit panels	—

D 5020 LIGHTING AND BRANCH WIRING

Includes	Excludes
Branch wiring and devices for lighting fixtures	Underfloor raceways *(see section D 5090, Other Electrical Systems)*
Lighting fixtures	Exterior site lighting *(see section G 4020, Site Lighting)*
Branch wiring for devices and equipment connections	—
Devices	—
Exterior building lighting	—

UNIFORMAT ELECTRICAL TITLES —
D 50 ELECTRICAL *(cont.)*

D 5030 COMMUNICATIONS AND SECURITY

Includes	Excludes
Fire alarm systems	Other electrical systems *(see section D 5090, Other Electrical Systems)*
Call systems	—
Telephone systems	—
Local area networks	—
Public address and music systems	—
Intercommunication systems and paging	—
Clock and program systems	—
Television systems	—
Security systems	—

D 5090 OTHER ELECTRICAL SYSTEMS

Includes	Excludes
Emergency generators	Electric baseboard *(see section D 3050, Terminal and Package Units)*
UPS	Electric coils and duct heaters *(see section D 3040, Distribution Systems)*
Emergency lighting systems	Building automation and energy monitoring systems *(see section D 3060, Controls and Instrumentation)*
Power factor correction	Communications and security systems *(see section D 5030, Communications and Security)*
Lightning and grounding protection systems	—
Raceway systems	—
Power generation systems	—

UNIFORMAT ELECTRICAL TITLES — G 40 SITE ELECTRICAL UTILITIES

G 4010 ELECTRICAL DISTRIBUTION

Includes	Excludes
Substations	—
Overhead power distribution	—
Underground power distribution	—
Ductbanks	—
Grounding	—

G 4020 SITE LIGHTING

Includes	Excludes
Fixtures and transformers	—
Poles	—
Wiring conduits and ductbanks	—
Controls	—
Grounding	—

G 4030 SITE COMMUNICATIONS AND SECURITY

Includes	Excludes
Overhead and underground communications	—
Site security and alarm systems	—
Ductbanks	—
Grounding	—

G 4040 OTHER SITE ELECTRICAL UTILITIES

Includes	Excludes
Cathodic protection	—
Emergency power generation	—

COMMON UNIT CONVERSIONS

Convert From	Multiply By	To Obtain
Acres	0.4047	Hectares
Acres	43,560	Square Feet
Acres	0.0016	Square Miles
Acres	4,047	Square Meters
Acres	0.0041	Square Kilometers
Acres	4840	Square Yards
Acre-feet	43560	Cubic Feet
Acre-feet	1,233	Cubic Meters
Acre-feet	1,613	Cubic Yards
Acre-feet	325,900	Gallons (US)
Acre-inches	3,630	Cubic Feet
Acre-inches	102.79	Cubic Meters
Acre-inches	134.44	Cubic Yards
Acre-inches	27,154	Gallons (US)
Atmospheres	76	Centimeters of Mercury
Atmospheres	29.92	Inches of Mercury
Atmospheres	1,033	Centimeters of Water
Atmospheres	33.9	Feet of Water
Atmospheres	101.325	Kilopascals
Atmospheres	101,325	Pascals
Atmospheres	14.7	Pounds per Square Inch
British Thermal Units	252.16	Calories
British Thermal Units	778.17	Foot-Pounds
British Thermal Units	0.0004	Horsepower-hours

COMMON UNIT CONVERSIONS *(cont.)*

Convert From	Multiply By	To Obtain
British Thermal Units	1,055	Joules
British Thermal Units	0.252	Kilogram-calories
British Thermal Units	107.51	Kilogram-meters
British Thermal Units	0.0003	Kilowatt-Hours
Calories	0.00397	British Thermal Units
Calories	4.184	Joules
Calories	3.086	Foot-Pounds
Centimeters	10	Millimeters
Centimeters	0.01	Meters
Centimeters	0.394	Inches
Centimeters	0.033	Feet
Centimeters	0.011	Yards
Cubic Feet	28.32	Liters
Cubic Feet	0.0283	Cubic Meters
Cubic Feet	1,728	Cubic Inches
Cubic Feet	957.51	Fluid Ounces (US)
Cubic Feet	59.844	Pints (US)
Cubic Feet	29.922	Quarts (US)
Cubic Feet	7.481	Gallons (US)
Cubic Feet	0.037	Cubic Yards
Cubic Feet per Second	448.83	Gallons (US) per Minute
Cubic Feet per Second	26,930	Gallons (US) per Hour

COMMON UNIT CONVERSIONS *(cont.)*		
Convert From	**Multiply By**	**To Obtain**
Cubic Feet per Minute	0.125	Gallons (US) per Second
Cubic Feet per Minute	448.83	Gallons (US) per Hour
Cubic Feet per Hour	0.002	Gallons (US) per Second
Cubic Feet per Hour	0.125	Gallons (US) per Minute
Cubic Inch	0.016	Liters
Cubic Inch	1.64×10^{-5}	Cubic Meters
Cubic Inch	0.554	Fluid Ounces (US)
Cubic Inch	0.035	Pints (US)
Cubic Inch	0.017	Quarts (US)
Cubic Inch	0.004	Gallons (US)
Cubic Inch	0.001	Cubic Feet
Cubic Inch	2.14×10^{-5}	Cubic Yards
Cubic Meters	1,000	Liters
Cubic Meters	264.172	Gallons (US)
Cubic Meters	35.315	Cubic Feet
Cubic Meters	1.308	Cubic Yards
Cubic Yards	764.555	Liters
Cubic Yards	0.765	Cubic Meters
Cubic Yards	46,656	Cubic Inches
Cubic Yards	201.974	Gallons (US)
Cubic Yards	27	Cubic Feet
Degrees (angle)	0.01745	Radians

COMMON UNIT CONVERSIONS *(cont.)*		
Convert From	**Multiply By**	**To Obtain**
Degrees (angle)	0.00278	Circles
Degrees (angle)	60	Minutes
Feet	304.8	Millimeters
Feet	30.48	Centimeters
Feet	0.305	Meters
Feet	3.05×10^{-4}	Kilometers
Feet	12	Inches
Feet	0.333	Yards
Feet	1.89×10^{-4}	Miles (statute)
Feet	1.65×10^{-4}	Miles (nautical)
Feet of Air	0.0009	Feet of Mercury
Feet of Air	0.00122	Feet of Water
Feet of Air	0.00108	Inches of Mercury
Feet of Air	0.00053	Pounds per Square Inch
Feet of Mercury	30.48	Centimeters of Mercury
Feet of Mercury	13.6086	Feet of Water
Feet of Mercury	163.3	Inches of Water
Feet of Mercury	5.8938	Pounds per Square Inch
Feet of Water	0.0295	Atmospheres
Feet of Water	2.2419	Centimeters of Mercury
Feet of Water	0.8826	Inches of Mercury
Feet of Water	304.78	Kilograms per Square Meter
Feet of Water	2988.9	Pascals
Feet of Water	62.424	Pounds per Square Foot
Feet of Water	0.4335	Pounds per Square Inch

Convert From	Multiply By	To Obtain
Feet per Second	0.305	Meters per Second
Feet per Second	1.097	Kilometers per Hour
Feet per Second	0.592	Knots
Feet per Second	0.682	Miles (statute) per Hour
Foot-Pounds	0.00129	British Thermal Units
Foot-Pounds	1.356	Joules
Foot-Pounds	0.324	Calories
Foot-Pounds	3.766×10^{-7}	Kilowatt-Hours
Gallons (US)	3.785	Liters
Gallons (US)	0.00379	Cubic Meters
Gallons (US)	231	Cubic Inches
Gallons (US)	128	Fluid Ounces (US)
Gallons (US)	8	Pints (US)
Gallons (US)	4	Quarts (US)
Gallons (US)	0.134	Cubic Feet
Gallons (US)	4.95×10^{-3}	Cubic Yards
Gallons (US) per Second	8.021	Cubic Feet per Minute
Gallons (US) per Second	481.25	Cubic Feet per Hour
Gallons (US) per Minute	2.23×10^{-3}	Cubic Feet per Second
Gallons (US) per Minute	8.021	Cubic Feet per Hour
Grams	1,000	Milligrams

COMMON UNIT CONVERSIONS *(cont.)*

Convert From	Multiply By	To Obtain
Grams	0.001	Kilograms
Grams	0.0353	Ounces
Grams	2.20×10^{-3}	Pounds
Hectares	10,000	Square Meters
Hectares	0.01	Square Kilometers
Hectares	107,639	Square Feet
Hectares	11,960	Square Yards
Hectares	2.471	Acres
Hectares	3.86×10^{-3}	Square Miles
Horsepower (US)	42.375	BTU per Minute
Horsepower (US)	2,543	BTU per Hour
Horsepower (US)	550	Foot-Pounds per Second
Horsepower (US)	33,000	Foot-Pounds per Minute
Horsepower (US)	1.014	Horsepower (Metric)
Horsepower (US)	0.7457	Kilowatts
Horsepower (US)	745.7	Watts
Inches	25.4	Millimeters
Inches	2.54	Centimeters
Inches	0.0254	Meters
Inches	0.0833	Feet
Inches	0.0278	Yards
Inches of Mercury	0.0334	Atmospheres

COMMON UNIT CONVERSIONS (cont.)

Convert From	Multiply By	To Obtain
Inches of Mercury	1.133	Feet of Water
Inches of Mercury	3,386	Pascals
Inches of Mercury	70.526	Pounds per Square Foot
Inches of Mercury	0.4912	Pounds per Square Inch
Inches of Water	2.46×10^{-3}	Atmospheres
Inches of Water	0.07355	Inches of Mercury
Inches of Water	25.398	Kilograms per Square Meter
Inches of Water	5.202	Pounds per Square Foot
Inches of Water	0.036	Pounds per Square Inch
Joules	9.5×10^{-4}	British Thermal Units
Joules	0.239	Calories
Joules	2.778×10^{-7}	Kilowatt-Hours
Joules	0.738	Foot-Pounds
Kilograms	1,000	Grams
Kilograms	35.274	Ounces
Kilograms	2.205	Pounds
Kilograms	1.1×10^{-3}	Short Tons
Kilograms	9.8×10^{-4}	Long Tons
Kilogram-calories	3.968	British Thermal Units
Kilograms per Square Meter	3.28×10^{-3}	Feet of Water
Kilograms per Square Meter	0.2048	Pounds per Square Foot
Kilograms per Square Meter	1.42×10^{-3}	Pounds per Square Inch

COMMON UNIT CONVERSIONS *(cont.)*		
Convert From	**Multiply By**	**To Obtain**
Kilometers	1,000	Meters
Kilometers	3,281	Feet
Kilometers	1,094	Yards
Kilometers	0.621	Miles (statute)
Kilometers per Hour	0.278	Meters per Second
Kilometers per Hour	0.54	Knots
Kilometers per Hour	0.911	Feet per Second
Kilometers per Hour	0.621	Miles (statute) per Hour
Kilopascals	9.87×10^{-3}	Atmospheres
Kilopascals	0.2952	Inches of Mercury
Kilopascals	4.021	Inches of Water
Kilopascals	1,000	Pascals
Kilopascals	20.885	Pounds per Square Foot
Kilopascals	0.145	Pounds per Square Inch
Kilowatts	56.8725	BTU per Minute
Kilowatts	1.341	Horsepower
Kilowatt-Hours	3,412	British Thermal Units
Kilowatt-Hours	3,600,000	Joules
Kilowatt-Hours	860,421	Calories
Kilowatt-Hours	2,655,000	Foot-Pounds
Knots	0.514	Meters per Second
Knots	1.852	Kilometers per Hour
Knots	1.688	Feet per Second
Knots	1.151	Miles (statute) per Hour

COMMON UNIT CONVERSIONS (cont.)

Convert From	Multiply By	To Obtain
Liters	0.001	Cubic Meters
Liters	61.024	Cubic Inches
Liters	33.814	Fluid Ounces (US)
Liters	0.264	Gallons (US)
Liters	0.0353	Cubic Feet
Liters	1.31×10^{-3}	Cubic Yards
Megapascals	145	Pounds per Square Inch
Meters	1,000	Millimeters
Meters	100	Centimeters
Meters	0.001	Kilometers
Meters	39.37	Inches
Meters	3.281	Feet
Meters	1.094	Yards
Meters	6.21×10^{-4}	Miles (statute)
Meters per Second	3.6	Kilometers per Hour
Meters per Second	1.944	Knots
Meters per Second	3.281	Feet per Second
Meters per Second	2.237	Miles (statute) per Hour
Miles (nautical)	1.1516	Miles (statute)
Miles (statute)	1,609	Meters
Miles (statute)	1.609	Kilometers
Miles (statute)	5,280	Feet
Miles (statute)	1,760	Yards
Miles (statute)	0.8684	Miles (nautical)

COMMON UNIT CONVERSIONS *(cont.)*

Convert From	Multiply By	To Obtain
Miles (statute) per Hour	0.447	Meters per Second
Miles (statute) per Hour	1.609	Kilometers per Hour
Miles (statute) per Hour	0.869	Knots
Miles (statute) per Hour	1.467	Feet per Second
Millimeters	0.1	Centimeters
Millimeters	0.001	Meters
Millimeters	0.0394	Inches
Millimeters	3.28×10^{-3}	Feet
Ounces	28,350	Milligrams
Ounces	28.35	Grams
Ounces	0.0283	Kilograms
Ounces	0.0625	Pounds
Ounces (Fluid, US)	0.0296	Liters
Ounces (Fluid, US)	1.805	Cubic Inches
Ounces (Fluid, US)	7.81×10^{-3}	Gallons (US)
Ounces (Fluid, US)	1.04×10^{-3}	Cubic Feet
Pascals	9.87×10^{-6}	Atmospheres
Pascals	3.35×10^{-4}	Feet of Water
Pascals	2.95×10^{-4}	Inches of Mercury
Pascals	4.01×10^{-3}	Atmospheres

COMMON UNIT CONVERSIONS *(cont.)*		
Convert From	**Multiply By**	**To Obtain**
Pascals	0.102	Kilograms per Square Meter
Pascals	0.021	Pounds per Square Foot
Pascals	1.45×10^{-4}	Pounds per Square Inch
Pounds	453.592	Grams
Pounds	0.454	Kilograms
Pounds	4.54×10^{-4}	Tons (Metric)
Pounds	16	Ounces
Pounds	0.0005	Tons (Short)
Pounds	4.46×10^{-4}	Tons (Long)
Pounds per Cubic Foot	16.02	Kilograms per Cubic Meter
Pounds per Cubic Foot	5.79×10^{-4}	Pounds per Cubic Inch
Pounds per Cubic Inch	27.68	Grams per Cubic Centimeter
Pounds per Cubic Inch	27,680	Kilograms per Cubic Meter
Pounds per Cubic Inch	1,728	Pounds per Cubic Foot
Pounds per Square Foot	0.016	Feet of Water
Pounds per Square Foot	4.89	Kilograms per Square Meter
Pounds per Square Foot	0.007	Pounds per Square Inch

COMMON UNIT CONVERSIONS *(cont.)*

Convert From	Multiply By	To Obtain
Pounds per Square Foot	47.88	Pascals
Pounds per Square Foot	0.048	Kilopascals
Pounds per Square Inch	2.31	Feet of Water
Pounds per Square Inch	27.73	Inches of Water
Pounds per Square Inch	0.0703	Kilograms per Square Centimeter
Pounds per Square Inch	2.036	Inches of Mercury
Pounds per Square Inch	144	Pounds per Square Foot
Pounds per Square Inch	6,895	Pascals
Pounds per Square Inch	6.895	Kilopascals
Pounds Per Square Inch	0.0069	Megapascals
Radians	57.3	Degrees
Square Centimeters	100	Square Millimeters
Square Centimeters	0.0001	Square Meters
Square Centimeters	0.155	Square Inches
Square Centimeters	1.08×10^{-3}	Square Feet
Square Centimeters	1.2×10^{-4}	Square Yards

COMMON UNIT CONVERSIONS *(cont.)*

Convert From	Multiply By	To Obtain
Square Feet	0.0929	Square Meters
Square Feet	9.3×10^{-6}	Hectares
Square Feet	9.3×10^{-8}	Square Kilometers
Square Feet	144	Square Inches
Square Feet	0.111	Square Yards
Square Feet	2.3×10^{-5}	Acres
Square Inches	645.16	Square Millimeters
Square Inches	6.452	Square Centimeters
Square Inches	6.45×10^{-4}	Square Meters
Square Inches	6.94×10^{-3}	Square Feet
Square Inches	7.72×10^{-4}	Square Yards
Square Kilometers	1,000,000	Square Meters
Square Kilometers	100	Hectare
Square Kilometers	10,760,000	Square Feet
Square Kilometers	1,196,000	Square Yards
Square Kilometers	247.105	Acres
Square Kilometers	0.386	Square Miles
Square Meters	10,000	Square Centimeters
Square Meters	0.0001	Hectares
Square Meters	0.000001	Square Kilometers
Square Meters	1,550	Square Inches
Square Meters	10.764	Square Feet
Square Meters	1.196	Square Yards
Square Meters	2.47×10^{-4}	Acres

COMMON UNIT CONVERSIONS *(cont.)*		
Convert From	**Multiply By**	**To Obtain**
Square Miles	2,590,000	Square Meters
Square Miles	259	Hectares
Square Miles	2.59	Square Kilometers
Square Miles	27,880,000	Square Feet
Square Miles	3,098,000	Square Yards
Square Miles	640	Acres
Square Millimeters	0.01	Square Centimeters
Square Millimeters	0.000001	Square Meters
Square Millimeters	1.55×10^{-3}	Square Inches
Square Millimeters	1.08×10^{-5}	Square Feet
Square Yards	0.836	Square Meters
Square Yards	8.36×10^{-5}	Hectares
Square Yards	8.36×10^{-7}	Square Kilometers
Square Yards	1296	Square Inches
Square Yards	9	Square Feet
Square Yards	2.07×10^{-4}	Acres
Tons (Metric)	1,000	Kilograms
Tons (Metric)	2,205	Pounds
Tons (Metric)	1.102	Tons (Short)
Tons (Metric)	0.984	Tons (Long)
Tons (Short)	907.184	Kilograms
Tons (Short)	0.907	Tons (Metric)
Tons (Short)	2000	Pounds
Tons (Short)	0.893	Tons (Long)

COMMON UNIT CONVERSIONS (cont.)

Convert From	Multiply By	To Obtain
Tons (Long)	1,016	Kilograms
Tons (Long)	1.016	Tons (Metric)
Tons (Long)	2,240	Pounds
Tons (Long)	1.12	Tons (Short)
Watts	3.4121	BTU per Hour
Watts	0.0568	BTU per Minute
Watts	0.0013	Horsepower
Watts	14.34	Calories per Minute
Watt-hours	3.4144	British Thermal Units
Yards	914.4	Millimeters
Yards	91.44	Centimeters
Yards	0.914	Meters
Yards	9.14×10^{-4}	Kilometers
Yards	36	Inches
Yards	3	Feet
Yards	5.68×10^{-4}	Miles (statute)

CONVERTING INCHES TO DECIMALS

Inches	Inches in Decimals	Feet In Decimals	Millimeters	Meters
1⁄16	0.0625	0.0052	1.5875	0.0016
1⁄8	0.1250	0.0104	3.1750	0.0032
3⁄16	0.1875	0.0156	4.7625	0.0048
1⁄4	0.2500	0.0208	6.3500	0.0064
5⁄16	0.3125	0.0260	7.9375	0.0079
3⁄8	0.3750	0.0313	9.5250	0.0095
7⁄16	0.4375	0.0365	11.1125	0.0111
1⁄2	0.5000	0.0417	12.7000	0.0127
9⁄16	0.5625	0.0469	14.2875	0.0143
5⁄8	0.6250	0.0521	15.8750	0.0159
11⁄16	0.6875	0.0573	17.4625	0.0175
3⁄4	0.7500	0.0625	19.0500	0.0191
13⁄16	0.8125	0.0677	20.6375	0.0206
7⁄8	0.8750	0.0729	22.2250	0.0222
15⁄16	0.9375	0.0781	23.8125	0.0238
1	1.0000	0.0833	25.4000	0.0254
2	2.0000	0.1667	50.8000	0.0508
3	3.0000	0.2500	76.2000	0.0762
4	4.0000	0.3333	101.6000	0.1016
5	5.0000	0.4167	127.0000	0.1270
6	6.0000	0.5000	152.4000	0.1524
7	7.0000	0.5833	177.8000	0.1778
8	8.0000	0.6667	203.2000	0.2032
9	9.0000	0.7500	228.6000	0.2286
10	10.0000	0.8333	254.0000	0.2540
11	11.0000	0.9167	279.4000	0.2794
12	12.0000	1.0000	304.8000	0.3048

OHM'S LAW

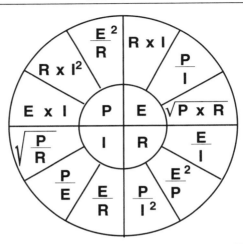

COMMON ELECTRICAL QUANTITIES

Symbol	Name	Unit of Measure
E	Voltage	Volt (V)
I	Current	Ampere (A)
R	Resistance	Ohm (Ω)
P	Power	Watt (W)
P	Power (apparent)	Volt-amp (VA)
C	Capacitance	Farad (F)
L	Inductance	Henry (H)
Z	Impedance	Ohm (Ω)
G	Conductance	Siemens (S)
f	Frequency	Hertz (Hz)
T	Period	Second (s)

ELECTRICAL FORMULAS

I = amperes	HP = horsepower
E = volts	% Eff. = percent efficiency
kW = kilowatts	PF = power factor
kVA = kilovolt/amperes	

To Find	Single Pass	Three Phase	Direct Current
AMPERES when kVA is known	$\dfrac{kVA \times 1000}{E}$	$\dfrac{kVA \times 1000}{E \times 1.73}$	Not Applicable
AMPERES when horsepower is known	$\dfrac{HP \times 746}{E \times \% \text{ Eff.} \times PF}$	$\dfrac{HP \times 746}{E \times 1.73 \times \% \text{ Eff.} \times PF}$	$\dfrac{HP \times 746}{E \times \% \text{ Eff.}}$
AMPERES when kilowatts are known	$\dfrac{kW \times 1000}{E \times PF}$	$\dfrac{kW \times 1000}{E \times 1.73 \times PF}$	$\dfrac{kW \times 1000}{E}$
KILOWATTS	$\dfrac{I \times E \times PF}{1000}$	$\dfrac{I \times E \times 1.73 \times PF}{1000}$	$\dfrac{I \times E}{1000}$
KILOVOLT/ AMPERES	$\dfrac{I \times E}{1000}$	$\dfrac{I \times E \times 1.73}{1000}$	Not Applicable
HORSE-POWER	$\dfrac{I \times E \times \% \text{ Eff.} \times PF}{746}$	$\dfrac{I \times E \times 1.73 \times \% \text{ Eff.} \times PF}{746}$	$\dfrac{I \times E \times \% \text{ Eff.}}{746}$
WATTS	$E \times I \times PF$	$E \times I \times 1.73 \times PF$	$E \times I$

WIRE SIZES

AWG gauge	Diameter Inches	Diameter mm	Ohms per 1000 ft.	Max. amps for chassis wiring	Max. amps for power trans-mission	Maximum frequency for 100% skin depth for solid conductor copper
OOOO	0.46	11.684	0.049	380	302	125 Hz
OOO	0.4096	10.40384	0.0618	328	239	160 Hz
OO	0.3648	9.26592	0.0779	283	190	200 Hz
0	0.3249	8.25246	0.0983	245	150	250 Hz
1	0.2893	7.34822	0.1239	211	119	325 Hz
2	0.2576	6.54304	0.1563	181	94	410 Hz
3	0.2294	5.82676	0.197	158	75	500 Hz
4	0.2043	5.18922	0.2485	135	60	650 Hz
5	0.1819	4.62026	0.3133	118	47	810 Hz
6	0.162	4.1148	0.3951	101	37	1100 Hz
7	0.1443	3.66522	0.4982	89	30	1300 Hz
8	0.1285	3.2639	0.6282	73	24	1650 Hz
9	0.1144	2.90576	0.7921	64	19	2050 Hz
10	0.1019	2.58826	0.9989	55	15	2600 Hz
11	0.0907	2.30378	1.26	47	12	3200 Hz
12	0.0808	2.05232	1.588	41	9.3	4150 Hz
13	0.072	1.8288	2.003	35	7.4	5300 Hz
14	0.0641	1.62814	2.525	32	5.9	6700 Hz
15	0.0571	1.45034	3.184	28	4.7	8250 Hz
16	0.0508	1.29032	4.016	22	3.7	11 kHz
17	0.0453	1.15062	5.064	19	2.9	13 kHz
18	0.0403	1.02362	6.385	16	2.3	17 kHz
19	0.0359	0.91186	8.051	14	1.8	21 kHz
20	0.032	0.8128	10.15	11	1.5	27 kHz
21	0.0285	0.7239	12.8	9	1.2	33 kHz
22	0.0254	0.64516	16.14	7	0.92	42 kHz
23	0.0226	0.57404	20.36	4.7	0.729	53 kHz

WIRE SIZES (cont.)

AWG gauge	Diameter Inches	Diameter mm	Ohms per 1000 ft.	Max. amps for chassis wiring	Max. amps for power transmission	Maximum frequency for 100% skin depth for solid conductor copper
24	0.0201	0.51054	25.67	3.5	0.577	68 kHz
25	0.0179	0.45466	32.37	2.7	0.457	85 kHz
26	0.0159	0.40386	40.81	2.2	0.361	107 kHz
27	0.0142	0.36068	51.47	1.7	0.288	130 kHz
28	0.0126	0.32004	64.9	1.4	0.226	170 kHz
29	0.0113	0.28702	81.83	1.2	0.182	210 kHz
30	0.01	0.254	103.2	0.86	0.142	270 kHz
31	0.0089	0.22606	130.1	0.7	0.113	340 kHz
32	0.008	0.2032	164.1	0.53	0.091	430 kHz
Metric 2.0	0.00787	0.200	169.39	0.51	0.088	440 kHz
33	0.0071	0.18034	206.9	0.43	0.072	540 kHz
Metric 1.8	0.00709	0.180	207.5	0.43	0.072	540 kHz
34	0.0063	0.16002	260.9	0.33	0.056	690 kHz
Metric 1.6	0.0063	0.16002	260.9	0.33	0.056	690 kHz
35	0.0056	0.14224	329	0.27	0.044	870 kHz
Metric 1.4	.00551	.140	339	0.26	0.043	900 kHz
36	0.005	0.127	414.8	0.21	0.035	1100 kHz
Metric 1.25	.00492	0.125	428.2	0.20	0.034	1150 kHz
37	0.0045	0.1143	523.1	0.17	0.0289	1350 kHz
Metric 1.12	.00441	0.112	533.8	0.163	0.0277	1400 kHz
38	0.004	0.1016	659.6	0.13	0.0228	1750 kHz
Metric 1	.00394	0.1000	670.2	0.126	0.0225	1750 kHz
39	0.0035	0.0889	831.8	0.11	0.0175	2250 kHz
40	0.0031	0.07874	1049	0.09	0.0137	2900 kHz

WIRE CURRENT CAPACITY

WIRE TYPES

Group 1: UF, TW
Group 2: RH, RHW, THW, THWN, XHHW, THHW, USE, FEPW, ZW
Group 3: THWN-2, XHH, USE-2, TA, TBS, SA, THHW, SIS, RHH, THW-2, THHN, XHHW, RHW-2, XHHW-2, ZW-2, FEP, MI
Group 4: UF, TW
Group 5: RH, RHW, THW, THWN, XHHW, THHW
Group 6: THWN-2, XHH, USE-2, TA, TBS, SA, THHW, SIS, RHH, THW-2, THHN, XHHW, RHW-2, XHHW-2, ZW-2

3 WIRES IN CABLE — AMBIENT TEMP 86°F (30°C)

Fahrenheit	140°F	167°F	194°F	140°F	167°F	194°F
Celsius	60°C	75°C	90°C	60°C	75°C	90°C
AWG	Group 1	Group 2	Group 3	Group 4	Group 5	Group 6
		COPPER			ALUMINUM	
18	—	—	14	—	—	—
16	—	—	18	—	—	—
14	20	20	25	—	—	—
12	25	25	30	20	20	25
10	30	35	40	25	30	35
8	40	50	55	30	40	45
6	55	65	75	40	50	60
4	70	85	95	55	65	75
3	85	100	110	65	75	85
2	95	115	130	75	90	100
1	110	130	150	85	100	115
1/0	125	150	170	100	120	135
2/0	145	175	195	115	135	150
3/0	165	200	225	130	155	175
4/0	195	230	260	150	180	205
250	215	255	290	170	205	230
300	240	285	320	190	230	255
350	260	310	350	210	250	280
400	280	335	380	225	270	305
500	320	380	430	260	310	350

WIRE CURRENT CAPACITY (cont.)

SINGLE WIRE IN CABLE — AMBIENT TEMP 86°F (30°C)

ahrenheit	140°F	167°F	194°F	140°F	167°F	194°F
Celsius	60°C	75°C	90°C	60°C	75°C	90°C
AWG	Group 1	Group 2	Group 3	Group 4	Group 5	Group 6
		COPPER			ALUMINUM	
18	—	—	18	—	—	—
16	—	—	24	—	—	—
14	25	30	35	—	—	—
12	30	35	40	25	30	35
10	40	50	55	35	40	40
8	60	70	80	45	55	60
6	80	95	105	60	75	80
4	105	125	140	80	100	110
3	120	145	165	95	115	130
2	140	170	190	110	135	150
1	165	195	220	130	155	175
1/0	195	230	260	150	180	205
2/0	225	265	300	175	210	235
3/0	260	310	350	200	240	275
4/0	300	360	405	235	280	315
250	340	405	455	265	315	355
300	375	445	505	290	350	395
350	420	505	570	330	395	445
400	455	545	615	355	425	480
500	515	620	700	405	485	545

ADJUSTMENTS TO WIRE CURRENT CAPACITY

AMBIENT TEMPERATURE ADJUSTMENT

For ambient temperature above 86°F (30°C), multiply the ampacities by the adjustment factor listed below for each group.

Fahrenheit	Celsius	Group 1	Group 2	Group 3
96-104	36-40	0.82	0.88	0.91
105-113	41-45	0.71	0.82	0.87
114-122	46-50	0.58	0.75	0.82
123-131	51-55	0.41	0.67	0.76
132-140	56-60	—	0.58	0.71
141-158	61-70	—	0.33	0.58
159-176	71-80	—	—	0.41

MORE THAN 3 WIRES IN CABLE ADJUSTMENT

For more than 3 wires in a cable, multiply the ampacities by the adjustment factor listed below by number of conductors.

Number of Conductors	Percentage
4-6	80%
7-9	70%
10-20	50%
21-30	45%
31-40	40%
41 and above	35%

CONDUIT SIZES

Trade Size (Inside Diameter)	Inches	Millimeters
1/2	0.622	15.8
3/4	0.824	20.9
1	1.049	26.6
1 1/4	1.38	35
1 1/2	1.61	40.9
2	2.067	52.5
2 1/2	2.469	62.7
3	3.068	77.9
3 1/2	3.548	90.1
4	4.026	102.3
5	5.047	128.2

CONDUIT WEIGHT TABLE
(POUNDS PER 100 FEET, EMPTY)

Size	Rigid Steel	IMC	Rigid Aluminum	EMT	PVC Schedule 40	PVC Coated Rigid Steel
1/2"	80	60	28	29	16	87
3/4"	108	82	37	45	22	115
1"	160	116	55	65	33	166
1 1/4"	208	150	72	96	46	217
1 1/2"	254	182	89	111	56	262
2"	344	242	119	141	74	367
2 1/2"	550	401	188	215	117	557
3"	710	493	246	260	153	724
3 1/2"	855	573	296	365	185	917
4"	1000	683	350	390	219	1056
5"	1335	—	479	—	298	1535
6"	1845	—	630	—	385	2025

SUPPORT SPACING FOR RIGID CONDUITS

RIGID METAL CONDUIT

Conduit Size	Max Distance Between Supports (ft.)
1/2" to 3/4"	10
1"	12
1 1/4" to 1 1/2"	14
2" to 2 1/2"	16
3" and larger	20

RIGID NON-METALLIC CONDUIT

Conduit Size	Max Distance Between Supports (ft.)
1/2" to 1"	3
1 1/4" to 2"	5
2 1/2" to 3"	6
3 1/2" to 5"	7
6" and larger	8

COPPER/ALUMINUM CONDUITS AND WIRES

Amps	Copper			Aluminum		
	Wire Size	Conduit Size	Ground	Wire Size	Conduit Size	Ground
20	#14	$1/2$"	#16	#12	$1/2$"	#12
25	#12	$1/2$"	#14	#10	$3/4$"	#12
35	#10	$3/4$"	#12	#8	$3/4$"	#10
50	#8	$3/4$"	#10	#6	1"	#8
65	#6	1"	#8	#4	$1 1/4$"	#6
85	#4	1"	#8	#2	$1 1/4$"	#6
100	#3	$1 1/4$"	#8	#1	$1 1/4$"	#6
115	#2	$1 1/4$"	#6	1/0	$1 1/2$"	#4
130	#1	$1 1/4$"	#6	2/0	2"	#4
150	1/0	$1 1/2$"	#6	3/0	2"	#4
175	2/0	$1 1/2$"	#6	4/0	$2 1/2$"	#4
200	3/0	2"	#6	250	$2 1/2$"	#4
230	4/0	2"	#4	300	$2 1/2$"	#2
255	250	$2 1/2$	#4	400	3"	#2
285	300	$2 1/2$"	#4	500	3"	#1
310	350	$2 1/2$"	#3	500	3"	#1
335	400	3"	#3	600	$3 1/2$"	#1
380	500	3"	#3	750	4"	#1
420	600	3"	#2	(2) 300	4"	1/0
460	700	$3 1/2$"	#2	(2) 350	5"	1/0
475	750	4"	#2	(2) 400	5"	1/0

EMT, IMC, RIGID CONDUIT FILL TABLES

Trade Size in Inches		Wire Size (THWN, THHN) Conductor Size AWG/kcmil																				
		14	12	10	8	6	4	3	2	1	1/0	2/0	3/0	4/0	250	300	350	400	500	600	700	750
1/2	EMT	12	9	5	3	2	1	1	1	1	1	–	–	–	–	–	–	–	–	–	–	–
	IMC	14	10	6	3	2	1	1	1	1	1	1	–	–	–	–	–	–	–	–	–	–
	Galv	13	9	6	3	2	1	1	1	1	1	–	–	–	–	–	–	–	–	–	–	–
3/4	EMT	22	16	10	6	4	2	1	1	1	1	1	1	1	–	–	–	–	–	–	–	–
	IMC	24	17	11	6	4	3	2	1	1	1	1	1	1	–	–	–	–	–	–	–	–
	Galv	22	16	10	6	4	2	1	1	1	1	1	1	1	–	–	–	–	–	–	–	–
1	EMT	35	26	16	9	7	4	3	3	1	1	1	1	1	–	–	–	–	–	–	–	–
	IMC	39	29	18	10	7	4	4	3	2	1	1	1	1	–	–	–	–	–	–	–	–
	Galv	36	26	17	9	7	4	3	3	1	1	1	1	1	–	–	–	–	–	–	–	–
1 1/4	EMT	61	45	28	16	12	7	6	5	4	3	2	1	1	1	1	1	1	–	–	–	–
	IMC	68	49	31	18	13	8	6	5	4	3	3	2	1	1	1	1	1	–	–	–	–
	Galv	63	46	29	16	12	7	6	5	4	3	2	1	1	1	1	1	1	–	–	–	–
1 1/2	EMT	84	61	38	22	16	10	8	7	5	4	3	3	1	1	1	1	1	1	1	1	1
	IMC	91	67	42	24	17	10	9	7	5	4	4	3	2	1	1	1	1	1	1	1	1
	Galv	85	62	39	22	16	10	8	7	5	4	3	3	2	1	1	1	1	1	1	1	1

	EMT	138	101	63	36	26	16	13	11	8	7	6	5	4	3	3	2	1	1	1	1	1
2	**IMC**	149	109	68	39	28	17	15	12	9	8	6	5	4	3	3	2	2	1	1	1	1
	Galv	140	102	64	37	27	16	14	11	8	7	6	5	4	3	3	2	2	1	1	1	1
	EMT	241	176	111	64	46	28	24	20	15	12	10	8	7	6	5	4	4	3	2	2	1
2 1/2	**IMC**	211	154	97	56	40	25	21	17	13	11	9	7	6	5	4	4	3	3	2	1	1
	Galv	200	146	92	53	38	23	20	17	12	10	8	7	6	5	4	3	3	2	1	1	1
	EMT	364	266	167	96	69	43	36	30	22	19	16	13	11	9	7	6	6	5	4	3	3
3	**IMC**	326	238	150	86	62	38	32	27	20	17	14	12	9	8	7	6	5	4	3	3	3
	Galv	309	225	142	82	59	36	31	26	19	16	13	11	9	7	6	5	5	4	3	3	3
	EMT	476	347	219	126	91	56	47	40	29	25	20	17	14	11	10	9	8	6	5	4	4
3 1/2	**IMC**	436	318	200	115	83	51	43	36	27	23	19	16	13	10	9	8	7	6	5	4	4
	Galv	412	301	189	109	79	48	41	34	25	21	18	15	12	10	8	7	7	5	4	4	4
	EMT	608	443	279	161	116	71	60	51	37	32	26	22	18	15	13	11	10	8	7	6	5
4	**IMC**	562	410	258	149	107	66	56	47	35	29	24	20	17	13	12	10	9	7	6	5	5
	Galv	531	387	244	140	101	62	53	44	33	27	23	19	16	13	11	10	8	7	6	5	5

PVC CONDUIT FILL TABLES

Trade Size in Inches		Wire Size (THWN, THHN) Conductor Size AWG/kcmil																					
		14	12	10	8	6	4	3	2	1	1/0	2/0	3/0	4/0	250	300	350	400	500	600	700	750
1/2	40	11	8	5	3	1	1	1	1	1	1	–	–	–	–	–	–	–	–	–	–	–
	80	9	6	4	2	1	1	1	1	1	1	–	–	–	–	–	–	–	–	–	–	–
3/4	40	21	15	9	5	4	2	1	1	1	1	1	1	1	–	–	–	–	–	–	–	–
	80	17	12	7	4	3	1	1	1	1	1	1	1	–	–	–	–	–	–	–	–	–
1	40	34	25	15	9	6	4	3	3	1	1	1	1	1	1	1	1	–	–	–	–	–
	80	28	20	13	7	5	3	3	2	1	1	1	1	1	1	1	–	–	–	–	–	–
1 1/4	40	60	43	27	16	11	7	6	5	3	3	2	1	1	1	1	1	1	1	1	1	1
	80	51	37	23	13	9	6	5	4	3	2	1	1	1	1	1	1	1	1	1	1	1
1 1/2	40	82	59	37	21	15	9	8	7	5	4	3	3	2	1	1	1	1	1	1	1	1
	80	70	51	32	18	13	8	7	6	4	3	3	2	1	1	1	2	1	1	1	1	1
2	40	135	99	62	36	26	16	13	11	8	6	6	5	4	3	3	2	1	1	1	1	1
	80	118	86	54	31	22	14	12	10	7	6	5	3	3	2	1	3	1	1	1	1	1
2 1/2	40	193	141	89	45	32	22	19	16	12	10	8	7	6	4	4	3	3	2	1	1	1
	80	170	124	78	45	32	20	17	14	10	9	7	6	5	4	3	3	3	2	1	1	1
3	40	299	218	137	79	57	35	30	25	18	15	13	11	9	7	6	5	5	4	3	3	2
	80	265	193	122	70	51	31	26	22	16	14	11	9	8	6	5	5	4	3	3	2	2
3 1/2	40	401	293	184	106	77	47	40	33	25	21	17	14	12	10	8	7	6	5	4	3	3
	80	358	261	164	95	68	42	35	30	22	18	15	13	10	8	7	6	6	5	4	3	3
4	40	517	377	238	137	99	61	51	43	32	27	22	18	15	12	11	9	8	7	6	5	4
	80	464	338	213	123	89	54	46	39	29	24	20	17	14	11	9	8	7	6	5	4	4
5	40	815	594	374	216	156	96	81	68	50	42	35	29	24	20	17	15	13	11	9	8	7
	80	736	537	338	195	141	86	73	61	45	38	32	26	22	18	15	13	12	10	8	8	7
6	40	1178	859	541	312	225	138	117	98	73	61	51	42	35	28	24	21	19	16	13	11	11

FLEXIBLE METALLIC CONDUIT FILL TABLES

Trade Size in Inches	Wire Size (THWN, THHN) Conductor Size AWG/kcmil																				
	14	12	10	8	6	4	3	2	1	1/0	2/0	3/0	4/0	250	300	350	400	500	600	700	750
1/2	13	9	6	3	2	1	1	1	1	1	—	—	—	—	—	—	—	—	—	—	—
3/4	22	16	10	6	4	2	1	1	1	1	1	1	1	—	—	—	—	—	—	—	—
1	33	24	15	9	6	4	3	3	1	1	1	1	1	1	1	1	—	—	—	—	—
1¹/₄	52	38	24	14	10	6	5	4	3	2	1	1	1	1	1	1	1	1	—	—	—
1¹/₂	76	56	35	20	14	9	7	6	4	4	3	2	1	1	1	1	1	1	1	1	1
2	134	98	62	35	25	16	13	11	8	7	6	5	4	3	3	2	1	1	1	1	2
2¹/₂	202	147	93	53	38	24	20	17	12	10	9	7	6	5	4	3	3	2	1	1	1
3	291	212	134	77	55	34	29	24	18	15	12	10	8	7	6	5	5	4	3	3	2
3¹/₂	396	289	182	105	76	46	39	33	24	20	17	14	12	9	8	7	6	5	4	4	3
4	518	378	238	137	99	61	51	43	32	27	22	18	15	12	11	9	8	7	5	5	4

LIQUID-TIGHT FLEXIBLE METALLIC CONDUIT FILL TABLES

Trade Size in Inches	Wire Size (THWN, THHN) Conductor Size AWG/kcmil																				
	14	12	10	8	6	4	3	2	1	1/0	2/0	3/0	4/0	250	300	350	400	500	600	700	750
1/2	13	9	6	3	2	1	1	1	1	1	—	—	—	—	—	—	—	—	—	—	—
3/4	22	16	10	6	4	2	1	1	1	1	1	1	1	—	—	—	—	—	—	—	—
1	36	26	16	9	7	4	3	3	1	1	1	1	1	1	1	1	—	—	—	—	—
1 1/4	63	46	29	16	12	7	6	5	4	3	2	1	1	1	1	1	1	1	1	1	1
1 1/2	81	59	37	21	15	9	8	7	5	4	3	3	2	1	1	1	1	1	1	1	1
2	133	97	61	35	25	15	13	11	8	7	6	5	4	3	3	2	1	1	1	1	1
2 1/2	201	146	92	53	38	23	20	17	12	10	8	7	6	5	4	3	3	2	1	1	1
3	308	225	141	81	59	36	30	26	19	16	13	11	9	7	6	5	5	4	3	3	3
3 1/2	401	292	184	106	76	47	40	33	25	21	17	14	12	10	8	7	6	5	4	4	3
4	523	381	240	138	100	61	52	44	32	27	23	19	15	12	11	9	8	7	6	5	5

METALLIC POWER RACEWAY FILL TABLES

500/700 RACEWAY

Cable/Wire Size	O.D.		500 Raceway	700 Raceway
	inches	[mm]	Number of Conductors	
14 AWG	0.111	[2.8]	7	10
12 AWG	0.130	[3.3]	5	7
10 AWG	0.164	[4.2]	3	4

2000 RACEWAY

Wire Size THHN/THWN	O.D.		Number of Conductors	
	inches	[mm]	Without Devices	Plugmold Receptacle
14 AWG	0.111	[2.8]	7	5
12 AWG	0.130	[3.8]	7	5

2100 RACEWAY

Wire Size THHN/THWN	O.D.		Number of Conductors		
	in.	[mm]	Without Devices	2127G Receptacle	2127GA GB GT Receptacle
14 AWG	0.111	[2.8]	34	4	9
12 AWG	0.130	[3.8]	25	3	6
10 AWG	0.164	[4.2]	15	2	4
8 AWG	0.216	[5.5]	9	0	0
6 AWG	0.254	[6.5]	6	0	0

METALLIC POWER RACEWAY FILL TABLES *(cont.)*

3000 RACEWAY

Wire Size THHN/THWN	O.D. inches	[mm]	Number of Conductors (40% Fill)				
			Without Devices	Duplex/Rect. Devices	Surge/GFCI Devices	Large Single	
14 AWG	0.111	[2.8]	152	70	40	28	
12 AWG	0.130	[3.8]	111	51	29	21	
10 AWG	0.164	[4.2]	70	32	18	13	
8 AWG	0.216	[5.5]	40	18	10	7	
6 AWG	0.254	[6.5]	29	13	7	5	

4000 RACEWAY

Wire Size THHN/THWN	O.D. inches	[mm]	Number of Conductors (40% Fill)							
			Without Devices		Duplex/Rect. Devices		Surge/GFCI Devices		Large Single Receptacles	
			Undivided	Divided	Undivided	Divided	Undivided	Divided	Undivided	Divided
14 AWG	0.111	[2.8]	296	127	165	78	107	49	49	20
12 AWG	0.130	[3.3]	216	93	120	57	78	36	36	15
10 AWG	0.164	[4.2]	136	58	76	36	49	22	22	9
8 AWG	0.216	[5.5]	78	33	43	20	28	13	13	5
6 AWG	0.254	[6.5]	56	24	31	15	20	9	9	3
4 AWG	0.324	[8.2]	34	15	0	0	0	0	0	0
3 AWG	0.352	[8.9]	29	12	0	0	0	0	0	0
2 AWG	0.384	[9.8]	24	10	0	0	0	0	0	0

METALLIC POWER RACEWAY FILL TABLES (cont.)

6000 RACEWAY

| Wire Size THHN/THWN | O.D. | | Number of Conductors (40% Fill) | | | | | | | | | |
| | | | Without Devices | | Duplex/Rect. Devices | | Surge/GFCI Devices | | Large Single Receptacles | |
	inches	[mm]	Undivided	Divided	Undivided	Divided	Undivided	Divided	Undivided	Divided
14 AWG	0.111	[2.8]	659	296	528	231	470	202	412	173
12 AWG	0.130	[3.3]	481	216	385	168	342	147	300	126
10 AWG	0.164	[4.2]	303	136	243	106	216	92	189	79
8 AWG	0.216	[5.5]	174	78	140	61	124	53	109	45
6 AWG	0.254	[6.5]	126	56	101	44	89	38	78	33
4 AWG	0.324	[8.2]	77	34	0	0	0	0	0	0
3 AWG	0.352	[8.9]	65	29	0	0	0	0	0	0
2 AWG	0.384	[9.8]	55	24	0	0	0	0	0	0
1 AWG	0.446	[11.3]	40	18	0	0	0	0	0	0
1/0 AWG	0.486	[12.3]	34	15	0	0	0	0	0	0
2/0 AWG	0.532	[13.5]	28	12	0	0	0	0	0	0

NON-METALLIC POWER RACEWAY FILL TABLES

400/800/2300 RACEWAY

Cable/ Wire Size	O.D. inches	O.D. [mm]	400 Raceway	800 Raceway	2300 Raceway
			Number of Conductors		
14 AWG	0.111	[2.8]	5	6	25
12 AWG	0.130	[3.3]	3	5	18
10 AWG	0.164	[4.2]	0	4	12

5000 RACEWAY

Wire Size THHN/THWN	O.D. inches	O.D. [mm]	Number of Conductors
14 AWG	0.111	[2.8]	24
12 AWG	0.130	[3.3]	20
10 AWG	0.164	[4.2]	12

NM 2000 RACEWAY

Wire Size THHN/THWN	O.D. inches	O.D. [mm]	Number of Conductors
14 AWG	0.111	[2.8]	45
12 AWG	0.130	[3.8]	31
10 AWG	0.164	[4.2]	15

BOX FILL TABLES

Number of Conductors in Outlet, Device, and Junction Boxes

Box Type	No. 18	No. 16	No. 14	No. 12	No. 10	No. 8	No. 6
4" × 1¹/₄" round or octagonal	8	7	6	5	5	4	2
4" × 1¹/₂" round or octagonal	10	8	7	6	6	5	3
4" × 2¹/₈" round or octagonal	14	12	10	9	8	7	4
4" × 1¹/₄" square	12	10	9	8	7	6	3
4" × 1¹/₂" square	14	12	10	9	8	7	4
4" × 2¹/₈" square	20	17	15	13	12	10	6
4¹¹/₁₆" × 1¹/₄" square	17	14	12	11	10	8	5
4¹¹/₁₆" × 1¹/₂" square	19	16	14	13	11	9	5
4¹¹/₁₆" × 2¹/₈" square	28	24	21	18	16	14	8
3" × 2" × 1¹/₂" device	5	4	3	3	3	2	1
3" × 2" × 2" device	6	5	5	4	4	3	2
3" × 2" × 2¹/₄" device	7	6	5	4	4	3	2
3" × 2" × 2¹/₂" device	8	7	6	5	5	4	2
3" × 2" × 2³/₄" device	9	8	7	6	5	4	2
3" × 2" × 3¹/₂" device	12	10	9	8	7	6	3
4" × 2¹/₈" × 1¹/₂" device	6	5	5	4	4	3	2
4" × 2¹/₈" × 1⁷/₈" device	8	7	6	5	5	4	2
4" × 2¹/₈" × 2¹/₈" device	9	8	7	6	5	4	2
3³/₄" × 2" × 2¹/₂" masonry box/gang	9	8	7	6	5	4	2
3³/₄" × 2" × 3¹/₂" masonry box/gang	14	12	10	9	8	7	2

MOTOR LOAD CURRENT			
SINGLE-PHASE MOTOR FULL LOAD CURRENT (AMPERES)			
Horsepower	115 V	208 V	230 V	Minimum Transformer kVA
1/6	4.4	2.4	2.2	0.53
1/4	5.8	3.2	2.9	0.70
1/3	7.2	4.0	3.6	0.87
1/2	9.8	5.4	4.9	1.18
3/4	13.8	7.6	6.9	1.66
1	16	8.8	8	1.92
1 1/2	20	11	10	2.40
2	24	13.2	12	2.88
3	34	18.7	17	4.10
5	56	30.8	28	6.72
7 1/2	80	44	40	9.60
10	100	55	50	12.00

Note:

1. If motors are started more than once per hour, add 20% additional kVA.

2. When motor service factor is greater than 1, increase full load amps proportionally (i.e., if service factor is 1.15, increase above values by 15%).

MOTOR LOAD CURRENT (cont.)

THREE-PHASE MOTOR FULL LOAD CURRENT (AMPERES)

Horsepower	208 V	230 V	460 V	575 V	Minimum Transformer kVA
1/2	2.2	2.0	1.0	0.8	0.9
3/4	3.1	2.8	1.4	1.1	1.2
1	4.0	3.6	1.8	1.4	1.5
1 1/2	5.7	5.2	2.6	2.1	2.1
2	7.5	6.8	3.4	2.7	2.7
3	10.7	9.6	4.8	3.9	3.8
5	16.7	15.2	7.6	6.1	6.3
7 1/2	24	22	11	9	9.2
10	31	28	14	11	11.2
15	46	42	21	17	16.6
20	59	54	27	22	21.6
25	75	68	34	27	26.6
30	88	80	40	32	32.4
40	114	104	52	41	43.2
50	143	130	65	52	52
60	170	154	77	62	64
75	211	192	96	77	80
100	273	248	124	99	103
125	342	312	156	125	130
150	396	360	180	144	150
200	528	480	240	192	200

Note:

1. If motors are started more than once per hour, add 20% additional kVA.

2. When motor service factor is greater than 1, increase full load amps proportionally (i.e., if service factor is 1.15, increase above values by 15%).

MOTOR LOAD CURRENT (cont.)

DIRECT-CURRENT MOTOR FULL LOAD CURRENT (AMPERES)

Horsepower	90 V	120 V	180 V	240 V	500 V	550 V
1/4	4.0	3.1	2.0	1.6	—	—
1/3	5.2	4.1	2.6	2.0	—	—
1/2	6.8	5.4	3.4	2.7	—	—
3/4	9.6	7.6	4.8	3.8	—	—
1	12.2	9.5	6.1	4.7	—	—
1 1/2	—	13.2	8.3	6.6	—	—
2	—	17.0	10.8	8.5	—	—
3	—	25.0	16.0	12.2	—	—
5	—	40.0	27.0	20.0	—	—
7 1/2	—	58.0	—	29.0	13.6	12.2
10	—	76.0	—	38.0	18.0	16.0
15	—	—	—	55.0	27.0	24.0
20	—	—	—	72.0	34.0	31.0
25	—	—	—	89.0	43.0	38.0
30	—	—	—	106.0	51.0	46.0
50	—	—	—	173.0	83.0	75.0
75	—	—	—	255.0	123.0	111.0
100	—	—	—	341.0	164.0	148.0

Note:

If motors are started more than once per hour, add 20% additional kVA.

TRANSFORMER LOAD CURRENT

SINGLE-PHASE TRANSFORMER FULL LOAD CURRENT (AMPERES)

kVA Rating	120 V	208 V	240 V	277 V	480 V	600 V
0.05	0.42	0.24	0.208	0.181	0.104	0.083
0.075	0.63	0.361	0.313	0.271	0.156	0.125
0.1	0.83	0.481	0.417	0.361	0.208	0.167
0.15	1.25	0.721	0.625	0.542	0.313	0.25
0.25	2.08	1.2	1.04	0.9	0.52	0.42
0.5	4.17	2.4	2.08	1.81	1.04	0.83
0.75	6.25	3.61	3.13	2.71	1.56	1.25
1	8.33	4.81	4.17	3.61	2.08	1.67
1.5	12.5	7.21	6.25	5.42	3.13	2.5
2	16.7	9.62	8.33	7.22	4.17	3.33
3	25	14.4	12.5	10.8	6.3	5
5	41.7	24	20.8	18.1	10.4	8.3
7.5	62.5	36.1	31.3	27.1	15.6	12.5
10	83.3	48.1	41.7	36.1	20.8	16.7
15	125	72.1	62.5	54.2	31.3	25
25	208	120.2	104.2	90.3	52.1	41.7
37.5	313	180.3	156.3	135.4	78.1	62.5
50	417	240.4	208.3	180.5	104.2	83.3
75	625	361	313	271	156	125
100	833	481	417	361	208	167
167	1392	803	696	603	348	278
200	1667	962	833	722	417	333
250	2083	1202	1042	903	521	417

Note:

When service factor is greater than 1, increase full load amps proportionally (i.e., if service factor is 1.15, increase above values by 15%).

TRANSFORMER LOAD CURRENT (cont.)

THREE-PHASE TRANSFORMER FULL LOAD CURRENT (AMPERES)

kVA Rating	208 V	240 V	480 V	600 V
3	8.33	7.22	3.61	2.89
6	16.7	14.4	7.2	5.8
9	25	21.7	10.8	8.7
15	41.6	36.1	18	14.4
22	61.1	53.0	26.5	21.6
30	83.3	72.2	36.1	28.9
45	124.9	108.3	54.1	43.3
75	208.2	180.4	90.2	72.2
112.5	312	271	135	108
150	416	361	180	144
225	625	541	271	217
300	833	722	361	289
500	1388	1203	601	481
750	2084	1806	903	723

Note:

When service factor is greater than 1, increase full load amps proportionally (i.e., if service factor is 1.15, increase above values by 15%).

TRANSFORMER WEIGHT (LBS) BY KVA

DRY TYPE 240/480 TO 120/240 V

1-Phase		3-Phase	
kVA	Lbs.	kVA	Lbs.
1	23	3	90
2	36	6	135
3	59	9	170
5	73	15	220
7.5	131	30	310
10	149	45	400
15	205	75	600
25	255	112.5	950
37.5	295	150	1140
50	340	225	1575
75	550	300	1870
100	670	500	2850
167	900	750	4300

OIL FILLED 3-PHASE 5/15 TO 480/277 V

kVA	Lbs.
150	1800
300	2900
500	4700
750	5300
1000	6200
1500	8400
2000	9700
3000	15000

GENERATOR WEIGHT (LBS) BY KW

GAS TYPE

3-Phase, 4-Wire 277/480 V

kW	Lbs.
7.5	600
10	630
15	960
30	1500
65	2350
85	2570
115	4310
170	6530

DIESEL TYPE

3-Phase, 4-Wire 277/480 V

kW	Lbs.
30	1800
50	2230
75	2250
100	3840
125	4030
150	5500
175	5650
200	5930
250	6320
300	7840
350	8220
400	10750
500	11900

VOLTAGE DROP TABLE

Formula: Voltage Drop = (Factor × Amps × Feet)/1000

Copper Conductor — 90% Power Factor

AWG	Single-Phase	3-Phase
14	0.4762	0.417
12	0.3125	0.263
10	0.1961	0.168
8	0.125	0.109
6	0.0833	0.071
4	0.0538	0.046
3	0.0431	0.038
2	0.0323	0.028
1	0.0323	0.028
0	0.0269	0.023
0	0.0222	0.02
0	0.019	0.016
0	0.0161	0.014
250	0.0147	0.013
300	0.0131	0.011
350	0.0121	0.011
400	0.0115	0.009
500	0.0101	0.009

LIGHT LEVELS FOR DIFFERENT PROJECT AREAS

The numbers listed below are average values. Please verify with specific project conditions as well as the local building codes.

Project Area	Footcandles
Accounting office	150
Auditorium	15
Bank lobby	50
Baseball field	100-150
Conference rooms	30-50
Corridors	10-20
Elevator	10
Factory assembly area	100-200
General office	50-80
Highway	1.5-2.0
Home kitchen	50-70
Hospital operating table	2500
Mechanical space	15
Parking lot	5
School classroom	60-90
Street	1.0-1.5
Warehouse	20-30

ESTIMATING EQUIPMENT PAD

CONCRETE VOLUME PER SQ. FT. OF SLAB

Thickness (in.)	Volume (CY)
4"	0.014
6"	0.021
8"	0.028
10"	0.035
12"	0.043
14"	0.050
16"	0.057
18"	0.064
20"	0.071
24"	0.085
30"	0.106

Estimating Example:

For an equipment pad of 12' long, 10' wide, 8" thick
Area of the pad: $10 \times 12 = 120$ SF
Concrete needed is about $120 \times 0.028 = 3.36$ CY

EXCAVATION SLOPES

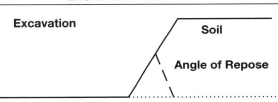

Types of Soil	Maximum Slopes (H:V)	Angle of Repose
Stable Rock	Vertical	90°
Type A (Hard and solid soil)	3/4:1	53°
Type B (Soil likely to crack or crumble)	1:1	45°
Type C (Soft, sandy, filled or loose soil)	1 1/2:1	34°

Note:

1. Excavation slopes are calculated by dividing horizontal distances by vertical distances. This definition of "excavation slope" is different from as "slopes" in other trades.

2. Data above only apply to excavations less than 20 feet deep.

3. Very few soils are stable enough for "Type A" or better. When no information is available, it is recommended to assume the soil to be Type C.

ESTIMATING TRENCH EXCAVATION

Total cubic yard (CY) of excavation required per linear foot (LF) of trench:

FOR EXCAVATION SLOPE OF VERTICAL (OR ANGLE OF REPOSE 90°)

Trench Width (LF)	Trench Depth (LF)								
	2	3	4	5	6	7	8	9	10
1	0.07	0.11	0.15	0.19	0.22	0.26	0.30	0.33	0.37
2	0.15	0.22	0.30	0.37	0.44	0.52	0.59	0.67	0.74
3	0.22	0.33	0.44	0.56	0.67	0.78	0.89	1.00	1.11
4	0.30	0.44	0.59	0.74	0.89	1.04	1.19	1.33	1.48
5	0.37	0.56	0.74	0.93	1.11	1.30	1.48	1.67	1.85
6	0.44	0.67	0.89	1.11	1.33	1.56	1.78	2.00	2.22
7	0.52	0.78	1.04	1.30	1.56	1.81	2.07	2.33	2.59

FOR EXCAVATION SLOPE OF ¾:1 (OR ANGLE OF REPOSE 53°)

Trench Width (LF)	Trench Depth (LF)								
	2	3	4	5	6	7	8	9	10
1	0.19	0.36	0.59	0.88	1.22	1.62	2.07	2.58	3.15
2	0.26	0.47	0.74	1.06	1.44	1.88	2.37	2.92	3.52
3	0.33	0.58	0.89	1.25	1.67	2.14	2.67	3.25	3.89
4	0.41	0.69	1.04	1.44	1.89	2.40	2.96	3.58	4.26
5	0.48	0.81	1.19	1.62	2.11	2.66	3.26	3.92	4.63
6	0.56	0.92	1.33	1.81	2.33	2.92	3.56	4.25	5.00
7	0.63	1.03	1.48	1.99	2.56	3.18	3.85	4.58	5.37

ESTIMATING TRENCH EXCAVATION (cont.)

FOR EXCAVATION SLOPE OF 1:1 (OR ANGLE OF REPOSE 45°)

Trench Width (LF)	Trench Depth (LF)								
	2	3	4	5	6	7	8	9	10
1	0.22	0.44	0.74	1.11	1.56	2.07	2.67	3.33	4.07
2	0.30	0.56	0.89	1.30	1.78	2.33	2.96	3.67	4.44
3	0.37	0.67	1.04	1.48	2.00	2.59	3.26	4.00	4.81
4	0.44	0.78	1.19	1.67	2.22	2.85	3.56	4.33	5.19
5	0.52	0.89	1.33	1.85	2.44	3.11	3.85	4.67	5.56
6	0.59	1.00	1.48	2.04	2.67	3.37	4.15	5.00	5.93
7	0.67	1.11	1.63	2.22	2.89	3.63	4.44	5.33	6.30

FOR EXCAVATION SLOPE OF 1½:1 (OR ANGLE OF REPOSE 34°)

Trench Width (LF)	Trench Depth (LF)								
	2	3	4	5	6	7	8	9	10
1	0.30	0.61	1.04	1.57	2.22	2.98	3.85	4.83	5.93
2	0.37	0.72	1.19	1.76	2.44	3.24	4.15	5.17	6.30
3	0.44	0.83	1.33	1.94	2.67	3.50	4.44	5.50	6.67
4	0.52	0.94	1.48	2.13	2.89	3.76	4.74	5.83	7.04
5	0.59	1.06	1.63	2.31	3.11	4.02	5.04	6.17	7.41
6	0.67	1.17	1.78	2.50	3.33	4.28	5.33	6.50	7.78
7	0.74	1.28	1.93	2.69	3.56	4.54	5.63	6.83	8.15

Estimating Example: For a trench 100' long, 4' wide, 2' deep, excavation slope of 1:1, then excavation volume is: 100 × 0.44 = 44 CY

ESTIMATING CONCRETE FOR CONDUIT ENCASEMENT

Note: The following data tables list the cubic yards of concrete for 100 feet of trench with 3" of concrete cover on all sides.

CONDUIT SEPARATION 1"

# of Conduits	Conduit Diameter					
	2"	3"	3$1/2$"	4"	4$1/2$"	5"
2	2.4	3.0	3.3	3.6	3.9	4.2
3	3.1	3.9	4.2	4.7	5.1	5.5
4	3.3	4.2	4.6	5.0	5.5	6.0
6	4.1	5.3	5.9	6.5	7.1	7.8
8	4.9	6.4	7.2	7.9	8.7	9.6
9	5.1	6.7	7.4	8.3	9.2	10.0

CONDUIT SEPARATION 1$1/2$"

# of Conduits	Conduit Diameter					
	2"	3"	3$1/2$"	4"	4$1/2$"	5"
2	2.6	3.2	3.4	3.7	4.0	4.4
3	3.3	4.1	4.5	5.0	5.3	5.8
4	3.6	4.5	4.9	5.4	6.0	6.5
6	4.6	5.9	6.5	7.2	7.9	8.6
8	5.7	7.3	8.0	8.9	9.8	10.7
9	5.9	7.6	8.5	9.5	10.3	11.3

ESTIMATING CONCRETE FOR
CONDUIT ENCASEMENT *(cont.)*

Note: The following data tables list the cubic yards of concrete for 100 feet of trench with 3" of concrete cover on all sides.

CONDUIT SEPARATION 2"

# of Conduits	Conduit Diameter					
	2"	3"	3¹/₂"	4"	4¹/₂"	5"
2	2.7	3.3	3.5	3.8	4.2	4.5
3	3.5	4.4	4.8	5.2	5.6	6.1
4	3.9	4.9	5.4	5.9	6.4	7.0
6	5.2	6.5	7.2	7.9	8.7	9.4
8	6.4	8.1	9.0	9.9	10.9	11.8
9	6.8	8.7	9.6	10.6	12.6	12.7

CONDUIT SEPARATION 3"

# of Conduits	Conduit Diameter					
	2"	3"	3¹/₂"	4"	4¹/₂"	5"
2	2.9	3.5	3.8	4.2	4.5	4.8
3	4.4	4.9	5.3	5.8	6.2	6.7
4	4.6	5.7	6.2	6.8	7.4	7.9
6	6.4	7.8	8.6	9.5	10.3	11.1
8	8.1	10.0	11.1	12.1	13.2	14.2
9	8.9	10.9	12.0	13.1	14.2	15.5

CHAPTER 9
Glossary

ESTIMATING AND BIDDING GLOSSARY

A

Addendum/Addenda: A written document adding to, clarifying or changing bidding documents. An addendum is generally issued after bid documents have been made available to contractors, but prior to bid closing. It is part of contract documents.

Agreement: Specific documents setting forth the terms of the contracts.

Allowance: In bidding, money set aside in contracts for items that have not been selected and specified. Bidders are required to include allowances in their proposal and contract amounts. For example, an electrical allowance sets aside an amount of money to be spent on electrical fixtures.

Alterations: Partial construction work done within an existing structure without new building addition.

Alternates: Amounts stated in bids to be added or deducted from the base bid amount proposed for alternative materials and/or construction methods.

The owner is to decide whether an alternate should be incorporated into contract sum at the time of award.

Application for Payment: Contractor's written request for payment for completed work and/or for materials delivered on-site.

Approved Bidders List: A list of contractors who have met pre-qualification criteria set by an owner.

Approved Equal: A contract clause stating that material or products finally supplied or installed must be equal to originally specified and approved by the architect/engineer.

Architect/Engineer: The professional hired by an owner to provide design services.

As-Built Drawings: Drawings marked up to reflect changes made during construction process to contract drawings, showing the locations, sizes, and nature of the building. They are permanent records for future reference.

B

Bid Bond: A bond issued by a surety to ensure that, if the bid is accepted, the contractor will sign a contract in accordance with his proposal.

Bid Documents: Drawings, details, and specifications for a particular project.

Bid Form: A standard written form furnished to all bidders so they can submit their bids using the same format.

Bid Opening: The actual process of opening and tabulating bids submitted. It can be open (where bidders are allowed to attend) or closed (where bidders are not allowed to attend).

Bid Security: A bid bond or certified check, guaranteeing that the bidder will sign a contract, if offered, in accordance with his proposal.

Bid Shopping: Negotiations between general contractors (buyers) and trade contractors (sellers) to obtain lower prices before/after submitting prime contract proposals to owners.

Bid Tab: A summary sheet listing all bid prices from contractors or suppliers.

Bill of Material: A list of items or components used for fabrication, shipping, receiving and accounting purposes.

Budget Estimate: Sometimes called a "ballpark" estimate, it's an estimate based on incomplete information like schematic drawings.

Builder's Risk Insurance: A form of property insurance covering a project under construction.

C

CO ("Certificate of Occupancy") A certificate issued by the local municipality after all inspections completed and all fees paid. It is required before anyone can occupy the constructed facility.

Certificate of Payment: Statement by an architect informing the owner of the amount due to a contractor based on work completed or materials stored.

Change Order: A written document signed by owner and contractor authorizing a change in the work or an adjustment in the contract sum or the contract time.

Claim: A formal notice sent by a contractor to an owner requesting additional compensation or an extension of time.

Closed Bid: A process where only invited bidders are allowed to submit proposals

Commencement of Work: The date when a written notice to proceed is sent from an owner to a contractor.

ESTIMATING AND BIDDING GLOSSARY *(cont.)*

Commissioning: When a project is near completion, the constructed facility is put into use to see if it functions as designed.

Conditions of the Contract: General, Supplementary and Special Conditions of a construction contract.

Construction Budget: The target cost figure covering the construction phase of a project. It does not include cost of land, A/E fees, or consultant fees etc.

Construction Schedule: A graphic, tabular or narrative representation of project construction phases, showing activities and durations in sequential order.

Contingencies: Amounts in a project budget dedicated to specific cost areas where oversight is an inevitable problem.

Contract Documents: A term used to represent all agreements between owner and contractor, any general, supplementary or special conditions, working drawings and specifications, all addenda and post-award change orders, etc.

Contract Overrun: The cost difference between the original contract price and the final completed cost including all adjustments made by approved change orders.

Contract Sum: The total agreeable amount payable by an owner to a contractor for the performed work under contract documents.

Contractor's Option: A written clause in contract documents giving a contractor the option of choosing certain specified materials, methods or systems without changing the contract sum.

Cost Breakdown: A breakdown furnished by a contractor describing portions of the contract sum allocated for principal divisions of the work.

Cost Codes: A numbering system given to specific portions of work to control construction costs.

Cost Plus Contract: A form of contract under which a contractor is reimbursed for his direct and indirect costs and is paid a fee for his services. The fee is usually stated as a stipulated sum or as a percentage of cost.

Critical Path Method (CPM): A scheduling diagram drawn to show the tasks involved in constructing a project. Critical path refers to the continuous chain of activities from project start to project finish, whose durations cannot be exceeded in order to complete the project on time.

CSI: Construction Specification Institute

CSI Master Format: A system of numbers and titles for organizing construction information into a standard order or sequence.

D

Design-Build Construction: A contractor bids or negotiates to provide design and construction services for the entire construction project

Direct Costs: The costs directly attributed to a work-scope, such as labor, material, and subcontracts. It does not include indirect costs like office overhead.

Direct Labor Costs: Costs from direct labor hours, not including the add-on portions like overtime, insurances, and payroll taxes.

Direct Material Costs: Costs for buying materials, including purchase price, freight, and taxes.

Draw: The amount of progress billings on a contract that is currently available to a contractor.

F

Factory Mutual (FM): An insurance agency that has established strict construction guidelines as relates to fire and environmental hazards.

Fast-Track: The process for a contractor to start the construction work before plans and specifications are complete.

Final Acceptance: The owner's acceptance of a completed structure from a contractor.

Final Inspection: A final site review by an owner before issuing the final payment.

Final Payment: The last payment from an owner to a contractor for the entire unpaid balance of the contract sum as adjusted by any approved change orders.

Fixed Fee Contract: A contract with fixed amounts for overhead and profit for all work performed.

Fixed Price Contract: A contract with a set price for the work, same as "lump sum"

Force Account Work: Work done and paid for on an expended time and material basis.

G

Gantt Chart: A project schedule that shows start and finish dates, critical and non-critical activities, slack time, and relationships among activities.

General Conditions: A written portion of contract documents stipulating contractor's minimum acceptable performance requirements, i.e. American Institute of Architects (AIA) Document A201.

General Contractor (or Prime Contractor): A properly licensed individual or company who takes full responsibility for project completion, although he may enter into subcontracts with others for the performance of specific portions of the project. A general contractor assembles and submits a proposal for the work on a project and then contracts directly with the owner to construct the project.

Guaranteed Maximum Price (GMP): A form of contracts where a contractor guarantees a ceiling price to the owner for construction cost

H

Hard Costs: Costs directly associated with construction.

I

Indirect Costs: Costs for items and activities not directly related to constructing a structure but are necessary to complete the project, i.e. contractor's overhead expense.

Indemnification Clause: Provision in a contract in which one party agrees to be financially responsible for specified types of damages, claims, or losses.

Instruction to Bidders: A document stating the procedures to be followed by bidders

Invitation to Bid: An invitation to a selected list of contractors including bid information related to their trades.

J

Jobsite Overhead: Necessary on-site construction expenses, such as construction support costs, supervision, bonus labor, field personnel, and jobsite office expenses.

L

Letter of Intent (Notice of Award): A notice from an owner to a contractor stating that a contract will be awarded to him. It usually functions as a formal notice to proceed on the project.

Lien: The right to hold or sell an owner's property for payment of debts to contractors.

Lien Release: A written document from contractor to owner that releases the lien following its satisfaction.

Lien Waiver: A written statement from a contractor or material supplier giving up lien rights against an owner's property.

Life-Cycle Cost: The cost of purchasing, installing, owning, operating, and maintaining a construction element over the life of a facility.

Liquidated Damages: An agreed-to amount chargeable against a contractor as reimbursement for damages suffered by the owner because of the contractor's failure to meet his obligations (i.e. not meeting the deadline).

ESTIMATING AND BIDDING GLOSSARY (cont.)

Long-Lead Items: Construction material and equipment which take significant time in fabrication and delivery to jobsite from their purchase dates, e.g. structural steel, elevator etc.

Lump Sum Contract: A written contract between the owner and contractor wherein the owner "agrees the pay" the contractor a fixed sum of money as a total payment for completing a scope of work.

M

Manufacturer's Specifications: The written installation and maintenance instructions developed by the manufacturer of a product and have to be followed in order to maintain the product warranty.

Maximum Occupancy Load: The maximum number of people permitted in a room and is measured per foot for each width of exit door. The maximum is 50 per foot of exit.

N

Notice to Proceed: A notice from an owner directing a contractor to begin work on a project.

O

Open Bid: A bid where any qualified bidder or estimator is given access to project information and allowed to submit a proposal.

Owner-Builder: A term used to describe an owner who takes the role of general contractor to build a project.

P

Payment Bond: A written form of security from a surety company to the owner, on behalf of a contractor, guaranteeing payment to all persons providing labor, materials, equipment, or services in accordance with the contract.

Penalty Clause: A contract provision that provides for a reduction in payment to a contractor as a penalty for failing to meet deadlines or requirements of contract specifications.

Performance Bond: A written form of security from a surety company on behalf of contractor to the owner that guarantees the contractors' proper and timely completion of a project.

Performance Specifications: Written material containing the minimum acceptable quality standards necessary to complete a project.

Permit: Written authorization from local governments giving permission to construct or renovate a building. Types of permits include zoning/use permit, demolition permit, grading permit, septic permit, building permit, electrical permit, plumbing permit etc.

Phased Construction: Constructing a facility in separate phases. Each separate phase is a complete project in itself.

Plan Submittal: Submission of construction plans to the city or county in order to obtain a building permit.

Preliminary Drawings: The drawings that precede the final approved drawings.

Pre-Qualification: A screening process of perspective bidders where an owner gathers background information for selection purposes. Considerations include competence, integrity, reliability, responsiveness, bonding capacity, similar project experience etc.

Product Data: Detailed information provided by material and equipment suppliers showing that the item provided meets the requirements of contract documents.

Progress Payment: Periodical payments to a contractor, usually based on the amount of work completed or material stored. There may be a temporary "retainage" or "hold back" (5% of the total value completed) which are to be refunded at the end of the project.

Project Directory: A written list of names, addresses, telephones and fax numbers for all parties connected with a specific project, which includes Owner, Architect, Attorney, General Contractor, Civil Engineer, Structural Engineer, etc.

Project Manual: An organized book that contains General Conditions, Supplementary and Special Conditions, the Form of Contract, Addenda, Change Orders, Bidding Requirements, Proposal Forms and the Technical Specifications.

Project Representative: A qualified individual authorized by the owner to assist in the administration of a construction contract.

Proposal: A written offer from a bidder to the owner to perform the work per specific prices and terms.

Proposal Form: A standard written form furnished to all bidders to submit proposals. It requires signatures from the authorized bidding representatives.

Punch List: A list of defects prepared by the owner that need to be corrected by the contractor immediately.

Q

Qualification Statement: A written statement of the contractor's experience and qualifications during the contractor selection process.

ESTIMATING AND BIDDING GLOSSARY *(cont.)*

R

Red Line: Blueprints that reflect changes and are marked with red pencil.

Reimbursable Expenses: Expenses that are to be reimbursed by the owner.

Retainage: An amount withheld from each payment to the contractor per Owner-Contractor Agreement.

RFI (Request for Information): A written request from a contractor to the owner or architect for clarification or information about the contract documents.

RFP: Request for Proposal.

S

Schedule of Values: The breakdown of a lump sum price into smaller portions of identifiable construction elements, which can be evaluated for progress payment purposes.

Separate Contract: A contract between an owner and a contractor, other than the general contractor, for constructing a portion of a project.

Shop Drawings: Drawings that show how specific portions of the work will be fabricated and installed.

Soft Costs: Cost items in addition to direct construction costs. They generally include architectural and engineering fees, legal, permits and development fees, financing fees, leasing and real estate commissions, advertising and promotion etc.

Special Conditions: Amendments to the General Conditions that change standard requirements to unique requirements for a specific project.

Specifications: Detailed statements describing materials, quality and workmanship to be used in a specific project. Most specifications use the Construction Specification Institute (CSI) format.

Standard Details: A drawing or illustration sufficiently complete and detailed for use on other projects with minimum or no changes.

Start-Up: The period prior to owner occupancy when mechanical, electrical, and other systems are activated and the owner's operating and maintenance staff is instructed in their use.

Subcontractor Bond: A written document from a subcontractor given to the general contractor, guaranteeing performance of his contract and payment of all labor, materials, and service bills associated with his subcontract agreement.

Substantial Completion: The stage when the construction is sufficiently complete and in accordance with the contract documents, so the owner may occupy the facility for the intended use.

Substitution: A proposed replacement offered in lieu of and as being equivalent to a specified material or process.

Substructure: The supporting part of a structure, i.e., the foundation.

Superstructure: The part of a building above the foundation.

Supplemental Conditions: A written section of the contract documents supplementing or modifying the standard clauses of general conditions.

Stop Order: A formal, written notification to a contractor to discontinue work on a project for reasons such as safety violations, defective materials or workmanship, or cancellation of the contract.

Subcontractor: A contractor hired by the general contractor to engage in a particular trade such as steel erection.

T

Take Off: Figuring materials necessary to complete a job.

T&M (Time and Materials): A construction contract which specifies a price for different elements of the work such as cost per hour of labor, overhead, profit, etc.

TI'S (Tenant Improvements): The interior improvement to a building after its shell is complete. It usually includes finish floor coverings, ceilings, partitions, doors with frames and hardware, fire protection, HVAC duct work and controls, electrical lighting and ancillary systems etc. The cost of TI is generally paid by the tenant.

Trade Contractor: A contractor that specialized in specific elements of the overall project.

Transmittal: A written document used to identify what is being sent to a receiving party. The transmittal is usually the cover sheet for the package sent and includes the contact information of the sender and receiver.

U

UL (Underwriters' Laboratories): An independent testing agency that checks products and materials for safety and quality.

Unit Price Contract: A written contract where an owner agrees to pay a contractor a specified amount of money for each unit of work completed. The designated unit price is all-inclusive: labor, materials, overhead, profit etc.

V

Value Engineering: A technical review process aimed to save construction costs by changing design options.

Verbal Quotation: A written document used by the contractor to record a subcontract or material cost proposal over the telephone, prior to receiving the written proposals via mail or fax later.

W

Warranty: There are two types of warranties. One is provided by the manufacturer of a product such as roofing material or an appliance. The other is a warranty for the labor. For example, a roofing contract may include a 20 year material warranty and a 5 year labor warranty.

Working Drawings: The actual drawings from which the building will be built.

Z

Zoning: A governmental process that limits the use of a property for a specific purpose, e.g. single family use, high rise residential use, industrial use, etc.

ELECTRICAL GLOSSARY

A

AC: abbreviation for alternating current, a current that continually reverses its direction. It is expressed in cycles per second (hertz or Hz).

A/D: Analog-to-digital conversion. The process changes an analog signal into a digital value representative of the magnitude of the signal at the moment of conversion.

Alternator: an electric generator designed to produce alternating current. Usually consists of rotating parts which created the changing magnetic field to produce the alternating current.

ANSI: American National Standards Institute, a private organization that coordinates and approves some US standards, particularly those related to the electrical industry.

Ampacity: the maximum continuous current that a conductor can carry without overheating above its temperature rating.

Ampere (Amp): A unit that measures the strength/rate of flow of electrical current. It is also equal to the flow of one coulomb per second. Named after French physicist Andre M. Ampère.

Ambient Temperature: the surrounding temperature.

Armored Cable: Electrical wires protected by metal sheathing

Ampere-Hour: the flow of electricity equal to one ampere for one hour. Commonly used to rate the capacity of batteries.

Analog: a measuring or display methodology which uses continuously varying physical parameters. In contrast, digital represents information in discrete binary form using only zeros and ones.

Anneal: Relief of mechanical stress through application of heat and gradual cooling. Annealing copper renders it soft and less brittle.

Apparent Power: The mathematical product of voltage and current on AC systems. Since voltage and current may not be in phase on AC systems, the apparent power thus calculated may not equal the real power, but may actually exceed it. Reactive loads (inductance and/or capacitance) on AC systems will cause the apparent power to be larger than the real power.

AWG: American Wire Gauge, a standard measure which represents the size of wire. The larger the number, the smaller the wire.

B

Battery: a group of two or more cells connected together to provide electrical current. Sometimes also used to describe a single cell which converts chemical energy to electrical current.

Battery Cycle Life: the number of discharge and recharge cycles that a battery can undergo before degrading below its capacity rating.

Battery Self-Discharge: the gradual loss of chemical energy in a battery that is not connected to any electrical load.

Baud Rate: A unit of measure for data transmission speed. It represents the number of signal elements (typically bits) transmitted per second. Typical baud rates are 300, 1200, 2400, 4800, 9600, 14400, and 28800.

Black Start: refers to certain electric utility generating units that can start upon demand without any outside source of electric power. These are often combustion turbines that have stationary battery banks to provide backup power to energize all the controls and auxiliaries necessary to get the unit up and running. In the event of a large area-wide blackout, these units are critical to restoring the utility grid. Most utility generating units do not have black start capability.

Bonding: an electrical conducting path formed by the permanent joining of metallic parts. Intended to assure electrical continuity and the capability to safely conduct any likely current. Similar to bonding jumper or bonding conductor

Branch Circuit: The circuits in a house that branch from the service panel to boxes and devices.

Breaker: A switch-like device that connects/disconnects power to a circuit.

Breakdown Voltage: The voltage at which the insulation between two conductors breaks down.

Bunch Stranding: A group of wires of the same diameter twisted together without a predetermined pattern.

Bus Bar: Separate, metallic strips that extend through the service panel. Breakers slide onto the "hot" busses and neutral and ground wires screw down in their respective busses.

C

Cabling: The twisting together of two or more insulated conductors to form a cable.

Cable Clamps: Metal clips inside an electrical box that hold wires in place.

Candela: the unit of luminous intensity.

Capacitor: a device that stores electrical charge usually by means of conducting plates or foil separated by a thin insulating layer of dielectric material. The effectiveness of the device, or its capacitance, is measured in Farads.

Capacitance: The ability of a dielectric material between conductors to store electricity when a difference of potential exists between the conductors. The unit of measurement is the farad, which is the capacitance value that will store a charge of one coulomb when a one-volt potential difference exists between the conductors. In AC, one farad is the capacitance value that will permit one ampere of current when the voltage across the capacitor changes at a rate of one volt per second.

Cell: a single device which converts chemical energy into electrical current. Sometimes also referred to as a battery.

Charge Rate: the rate at which a battery or cell is recharged. Can be expressed as a ratio of battery capacity to current flow.

Circuit: A continuous loop of current (i.e. incoming "hot" wire, through a device, and returned by "neutral" wire).

Circuit Breaker: The most common type of "over current protection." A breaker trips when

ELECTRICAL GLOSSARY (cont.)

a circuit becomes overloaded or shorts out.

Color Code: A system for circuit identification through use of solid colors and contrasting tracers.

Conductor: usually an uninsulated wire capable of transmitting electricity with little resistance. The best conductor at normal temperature ranges is silver. The most common is copper. Some other recently discovered substances called super conductors actually have zero resistance at extremely low temperatures.

Conduit: A protective metal tube that wires run through.

Contacts: Elements used to mechanically make or break an electric circuit.

Continuity Check: A test to determine whether electrical current flows continuously throughout the length of a single wire or individual wires in a cable.

Continuous Load: a sustained electrical load current for three hours or more.

Cord: A flexible insulated cable.

Corona: Ionization of air surrounding a conductor caused by the influence of high voltage. Causes deterioration of insulation materials.

Coulomb: the practical unit of electric charge transmitted by a

current of one ampere for one second. Named for the French physicist Charles A. de Coulomb.

Current: the flow of electricity commonly measured in amperes.

Current Carrying Capacity: The maximum current an insulated conductor can safely carry without exceeding its insulation and jacket temperature limitations. It is dependent on the installation conditions.

Cycles-Per-Second: A measure of the frequency in an AC electric system. Abbreviated CPS or cycles. Now replaced with the unit Hertz.

D

DC: Direct current, electrical current that normally flows in one direction only.

Decibel (db): A unit that expresses differences of power or voltage level. It is used to express power loss in passive circuits or cables.

Depth of Discharge: The percent of rated capacity of a battery that has been discharged from the battery.

Dielectric Strength: The voltage that an insulation can withstand before breakdown occurs. Usually expressed as a voltage gradient (such as volts per mil).

Diode: An electronic semiconductor device that

predominantly allows current to flow in only one direction.

Duplex Receptacle: The commonly used receptacle (outlet). Called "duplex" because it has two plug-in sockets.

E

Efficiency of a Light Source: The total light output of a light source divided by the total power input. Efficiency is expressed in lumens per Watt.

Electrolyte: A nonmetallic conductor of electricity usually consisting of a liquid or paste in which the flow of electricity is by ions.

Energy: The ability to do mechanical work. Electrical energy is measured in kilowatt-hours for billing purposes.

F

Factory Calibration: The tuning or altering of a control device by the manufacturer to bring it into specification.

Farad: A unit of electrical capacity. One coulomb of charge will produce a potential difference of one volt across a capacitance of one Farad. Named for the English physicist Michael Faraday.

Fault: A short circuit.

Feeder: Circuit conductors between the service equipment

and the last downstream branch circuit over current protective device.

Filter: A device made up of circuit elements designed to pass desirable frequencies and block all others. It typically consists of capacitors and inductors.

Fixture: Any permanently connected light or other electrical device that consumes power.

FLA: Full load amperes, the current that a motor requires to produce rated nameplate horsepower output when rated voltage and frequency is provided to it's terminals.

Float Charge: Charging current supplied to a battery which overcomes the battery self-discharge rate. This current, under otherwise normal conditions, will maintain the battery in a fully charged state.

Flex Life: The measurement of the ability of a conductor or cable to withstand repeated bending.

Footcandle: The amount of light produced by a lamp measured in lumens divided by the area that is illuminated.

Frequency: The number of times an alternating current repeats its cycle in one second. It is measured in Hertz. The standard frequency in the US is 60 Hz. However, in some other countries the standard is 50 Hz.

ELECTRICAL GLOSSARY (cont.)

Fuses: Removable devices that link a circuit at the fuse box. Fuse connections blow apart and break the circuit if an overload or short occurs.

G

Gain: Ratio of output voltage, current, or power to input voltage, current, or power.

Gassing: Gas by-products produced by the chemical reactions that occur when charging a battery. Since one of these gasses is often hydrogen, safety precautions must be taken to ensure proper ventilation to avoid the danger of explosion.

Generator: A rotating machine which converts mechanical energy into electrical energy.

GFCI or GFI (Ground Fault Circuit Interrupter): A specific type of circuit protection (commonly required in kitchens and bathrooms) that helps safeguard against shocks. GFCI protection can come from an outlet or a breaker.

Grid: A term used to refer to the electrical utility distribution network.

Ground: A conducting connection between an electrical circuit or device and the earth.

Grounding: An electrical system is grounded to provide a path of least resistance from the power source to the earth so voltage is drained away before it can do any damage to people or equipment. The NEC considers an electrical system grounded when it's "connected to earth or to some conducting body that serves in place of the earth."

Ground Fault: Current misdirected from the hot (or neutral) lead to a ground wire, box, or conductor.

H

Harmonic: A sine wave which is an integral multiple of a base frequency. For example, the third harmonic on a 60 Hz system is a frequency of 180 Hz. Certain types of electrical equipment generate harmonics which interfere with the proper functioning of other devices connected to the same system.

Harness: An arrangement of wires and cables, usually with many breakouts, which have been tied together or pulled into a rubber or plastic sheath, used to interconnect an electric circuit.

Henry: The practical unit of inductance. One Henry is equal to the inductance in which the change of one ampere per second results in an induced voltage of one volt. Abbreviated H. Named for the American physicist Joseph Henry.

ELECTRICAL GLOSSARY (cont.)

Hertz: Unit of frequency. One Hertz equals one complete cycle per second of an AC source. Abbreviated Hz. Named after the German physicist Heinrich R. Hertz.

Hi-Pot: A test designed to determine the highest voltage that can be applied to a conductor without breaking through the insulation.

Horsepower: A unit of power equal to 746 watts.

Hot, Neutral, Ground: The three most common circuit wires. The hot brings the current flow in, the neutral returns it to the source, and the ground is a safety route for returning current. The ground and neutral are joined only at the main service panel.

I

IEEE: Abbreviation for Institute of Electrical and Electronics Engineers, an independent organization which develops electrical standards and furthers the profession of electrical and electronics engineers.

Impedance: The total effects of a circuit that oppose the flow of an AC current consisting of inductance, capacitance, and resistance. It can be quantified in the units of ohms.

Inductance: The property of a circuit or circuit element that opposes a change in current flow, thus causing current changes to lag behind voltage changes. This can be a variation of the current in the circuit itself (self-inductance) or in a nearby circuit (mutual inductance). It is measured in Henrys.

Inductive Load: Electrical devices made of wound or coiled wire. Current passing through the coil creates a magnetic field that in turn produces mechanical work.

Insulation: A material having high resistance to the flow of electric current. Often called a dielectric in radio frequency cable.

Inverter: An electrical device which is designed to convert direct current into alternating current.

Ion: A positively or negatively charged atom or molecule.

J

Jacket: An outer non-metallic protective covering applied over an insulated wire or cable.

Joule: A unit of work or energy equal to one watt for one second. Named after James P. Joule, an English physicist. 1 Kwhr = 2,655,000 ft-lb = 1.341 hp-hr = 3413 Btu = 3,600,000 joules.

Jumper Cable: A short flat cable interconnecting two wiring boards or devices.

Junction (Electrical) Box: A square, octagonal, or rectangular plastic or metal box that fastens to framing and houses wires, and/or receptacles and/or switches.

K

Kilovar (kVAr or KVAR): Unit of AC reactive power equal to 1000 vars.

Kilovolt (kV or KV): Unit of electrical potential equal to 1000 volts.

Kilovolt amperes (kVA or KVA): a unit of apparent power equal to 1000 volt amperes.

Kilowatt (kW or KW): Unit of power equal to 1000 watts

Kilowatt-hour (kwh or KWH): Unit of energy or work equal to one kilowatt for one hour. This is the normal quantity used for metering and billing electricity customers. At a 100% conversion efficiency, one kwh is equivalent to about 4 fluid ounces of gasoline, $3/16$ pound LP, 3 cubic feet natural gas, or $1/4$ pound coal.

Knockout: A removable piece of an electrical box or panel that's "knocked out" to allow cable to enter the box

L

Lead: The short length of a conductor that hangs free in a box or service panel. (i.e. a wire end.)

Listed: an electrical device or material that has been tested by a recognized organization

and shown to meet appropriate standards. Many local governmental authorities require that installed electrical products be listed. A well-known listing organization is Underwriters Laboratories (UL).

Load: a device which consumes power and is connected to an electricity source.

LRA: Locked rotor amperes, the current that a motor would require if the rotor were locked in place and prevented from rotating and rated nameplate voltage and frequency were applied to its terminals. This LRA value is important when sizing a generator because the generator's surge rating must be large enough to handle it.

Lumen: The unit used to measure the total amount of light produced by a light source.

Luminance: Luminous Flux (light output). This is the quantity of light that leaves the lamp, measured in lumens (lm).

M

MCA: Minimum circuit amperes, the minimum current rating allowed for the wiring and circuit breaker or fuse protection for the equipment. It is used by the installer and electrician to size the branch circuit to feed the equipment.

Multi-Conductor: More than one conductor within a single cable complex.

N

NEC: National Electrical Code, a code containing information regarding wiring design and protection, wiring methods and materials, equipment, special occupancies etc. It is sponsored and regularly updated by the National Fire Protection Association.

Neutral: A conductor of an electrical system which usually operates with minimal voltage to ground. Depending on the type of system, it may carry little current or only unbalance current.

NM: Nonmetallic-sheathed (plastic).

NMC: Solid plastic nonmetallic-sheathing used in wet or corrosive areas, but not underground (see UF).

O

Ohm: A unit that measures the resistance a conductor has to electricity. A circuit resistance of one ohm will pass a current of one ampere with a potential difference of one volt. Abbreviated using the Greek letter omega. Named for the German physicist George Simon Ohm.

On/Off Control: A simple control system in which the device being controlled is either full on or full off, with no intermediate operating positions.

Open Circuit Voltage: The maximum voltage produced by a power source with no load connected.

OSHA: Occupational Safety and Health Act, specifically the Williams-Steiger Law passed in 1970 covering all factors relating to safety in places of employment.

Over Current: Any current beyond the continuous rated current of the conductor or equipment. This may be value slightly above the rating as in the case of an overload, or may be far above the rating as in the case of a short circuit.

Overload: Operation of electrical equipment above its normal full-load rating or of a conductor above its rated ampacity. An overload condition will eventually cause dangerous overheating and damage.

P

Parallel Transmission: The transmission of data bits over different lines, usually simultaneously; as opposed to serial transmission.

Pigtail: A short, added piece of wire connected by a wire connector. Commonly used to extend or connect wires in a box.

Potting: The sealing of a cable termination or other component with a liquid that cures into an elastomer.

Power: The rate at which work is performed or that energy is transferred. Electric power is commonly measured in watts or kilowatts. A power of 746 watts is equivalent to 1 horsepower.

Power Factor: The ratio of real power to apparent power delivered in an AC electrical system or load. Its value is always in the range of 0.0 to 1.0 or 0% to 100%. A unity power factor (1.0) indicates that the current is in phase with the voltage and that reactive power is zero.

R

Range: The limits within which a device or circuit operates or the distance over which a transmitter operates reliably.

Rated Voltage: The maximum voltage at which an electrical component can operate for extended periods without undue degradation or safety hazard.

REA: Rural Electrification Administration, which is part of the U.S. Dept. of Agriculture. REA establishes specifications and provides approval for telephone station wire and power cable.

Reactive Power: The mathematical product of voltage and current consumed by reactive loads. Examples of reactive loads include capacitors and inductors. These types of loads when connected to an AC voltage source will draw current, but they actually consume no real power in the ideal sense.

Real Power: The rate at which work is performed or that energy is transferred. Electric power is commonly measured in watts or kilowatts. The term real power is often used to differentiate from reactive power. Also called active power.

Resistance: The characteristic of materials to oppose the flow of electricity in an electric circuit. It is measured in ohms.

RLA: Running load amperes, also sometimes abbreviated FLA for full load amperes. This is the current in amperes that a motor requires to produce rated nameplate horsepower output when rated nameplate voltage and frequency is provided to it's terminals.

RMS: "Root-Mean-Square", a method of computing the effective value of a time-varying electrical wave. For example, an AC current is said to have an effective or RMS value of one ampere when it produces heat in a certain resistance at the same average rate as a continuous (or DC) current of one ampere would in the same resistance.

Romex: A brand name of nonmetallic-sheathed cable made by General Cable Corporation. Often mistakenly used as a collective term for NM sheathed cable

Rough-In: Installing the boxes, cables, and making "in-wall" connections while the walls are still open. Later, final connections are made and the devices and appliances are installed during the trim-out.

S

Separately Derived System: An electrical system whose power is provided by a stand-alone generator, transformer, or converter and which has no direct electrical connection or ground connection to another source (such as the utility). The NEC contains special grounding and bonding requirements for such systems.

Serial Transmission: Sending one data bit at a time on a single transmission line.

Series (Universal) Motor: A non-induction type motor utilized for small equipment. Speed will decrease as load increases.

Service: The equipment and conductors that transmit electricity from the utility supply system to the building being served.

Service Equipment: The circuit breaker or fused switch located near where the service conductors enter the building. They are intended as the primary means of disconnecting the supply.

Service Entrance (SE): The location where the incoming electrical line enters the home.

Service/Supply Leads: The incoming electrical lines that supply power to the service panel.

Service Panel: The main circuit breaker panel (or fuse box) where all the circuits tie into the incoming electrical supply line

Shield: A sheet, screen, or braid of metal, usually copper, aluminum, or other conducting material, placed around or between electric circuits or cables or their components to contain any unwanted radiation, or to keep out an unwanted interference.

Short Circuit: When current flows "short" of reaching a device. Caused by a hot conductor accidentally contacting a neutral or ground. A short circuit is an immediate fault to ground and should always cause the breaker to trip or the fuse to blow. (Also see ground fault)

Shunt: A conductor joining two points in an electrical circuit to form a parallel path. All or some portion of the current may pass through the shunt.

ELECTRICAL GLOSSARY (cont.)

Sine Wave: In ideal electric systems, the characteristic shape of the alternating voltage or current wave.

Single-Phase: An AC electric system or load consisting of at least one pair of conductors energized by a single alternating voltage. This type of system is simpler than three-phase but has substantial disadvantages when large amounts of power have to be delivered.

Single-Phase Motor: Any motor energized by a single alternation voltage.

Strand: A single uninsulated wire.

Stranded Conductor: A conductor composed of groups of wires twisted together.

Surge: A temporary and relatively large increase in the voltage or current in an electric circuit or cable. Also called transient.

Surge Capacity: The ability of an electrical supply to tolerate a momentary current surge or inrush imposed by the starting of motors or the energizing of transformers.

T

TC: Tray Cable, multi-conductor cable specifically approved for use installed in cable trays.

Three-Phase: An AC electric system or load consisting of three conductors energized by alternating voltages that are out of phase by one third of a cycle. This type of system has advantages over single-phase including the ability to deliver greater power using the same ampacity conductors and the fact that it provides a constant power throughout each cycle rather than a pulsating power, as in single-phase. Large power installations are three-phase.

Three-Phase Motor: A relatively inexpensive, self-starting motor (no starting winding or capacitor); can start heavy loads. The motor requires a three-phase AC power supply.

Tolerance: The maximum allowable deviation from a specified standard, as the range of variation permitted, expressed in actual values or more often as a percentage of the nominal value.

Transducer: Any device which generates an electrical signal from real world physical measurements.

Transformer: A device that converts one AC voltage and current to a different voltage and current. Constructed using two or more coils of wire around a common magnetic core. Transformers are an efficient and economical means of transferring large amounts of AC electric power at high voltages. This is the primary advantage of AC systems over DC systems.

ELECTRICAL GLOSSARY (cont.)

Transmitter: A device which translates the low-level output of a sensor or transducer to a higher level signal suitable for transmission to a site where it can be further processed.

Travelers: Wires that carry current between three-way and/or four-way switches.

U

UF (Underground Feeder) Cable: Cable designed and rated for underground, outdoor use. Cable wires are molded into solid plastic.

UL: Underwriters Laboratories, a non-profit organization that was established by the insurance industry to test electrical devices for safety.

UPS: Uninterruptible power supply, a device that provides a constant regulated voltage output in spite of interruptions of the normal power supply. It includes filtering circuits and is usually used to feed computers or related equipment which would otherwise shutdown on brief power interruptions.

V

VA: Volt ampere, a unit of apparent power equal to the mathematical product of a circuit voltage and amperes.

VAR: Volt ampere reactive, unit of AC reactive power.

Volt: A unit that measures the amount of electrical pressure. It is the potential that will produce a current of 1 ampere through a resistance of 1 ohm. Named after Italian physicist Count Alessandro Volta.

Voltage Drop: A voltage reduction due to impedances between the power source and the load. These impedances are due to wiring and transformers and are normally minimized to the extent possible.

W

Watt: A unit that measures the amount of electrical power. Named after the Scottish engineer James Watt

Water Resistant: UL designation for cords that have an insulation on the individual conductors that passes UL requirements (i.e. ST Water Resistant or ST Dry 105°C, Water Resistant 60°C).

CHAPTER 10
Abbreviations

ABBREVIATIONS

A	Area Square Feet, Ampere	**CB**	Circuit Breaker
AC	Alternating Current	**CCW**	Counter Clock Wise
ACC	Accessory	**CFM**	Cubic Feet per Minute
ACCU	Accumulator	**CIR**	Circuit
ACPTR	Acceptor	**CNDCT**	Conductor
ACT	Actuator	**CNDT**	Conduit
ACTI	Activated	**COMP**	Compressor
ADD'L	Additional	**COND**	Condenser
ADJ	Adjustable	**CONT**	Contactor
ADPT	Adapter	**CPC**	Circuit Protective Conductor
AL	Aluminum		
ALRM	Alarm	**CPS**	Cycles per Second
ALT	Altitude, Alternate	**CR**	Control Relay
ANNN	Annunciator	**CSA**	Cross-sectional Area
AP	Accessory Package	**CU**	Copper, Cubic
ASTM	American Society of Testing and Materials	**CW**	Clockwise
		CWT	100 Pounds
ATNT	Attentuator	**DBL**	Double
AWG	American Wire Gauge	**DC**	Direct Current
AWM	Appliance wiring material	**DDC**	Direct Digital Control
		DECO	Decorative
BAL	Balancing	**DEV**	Device
BC	Bayonet Cap	**DIAG**	Diagram
BFFL	Baffle	**DIM**	Dimming
BLCK	Block	**DISC**	Disconnect, Discharge
BPM	Blows per Minute		
BRD	Board	**DISP**	Disposable, Display
BRG	Bearing	**DIST**	Distribution
BRKR	Breaker	**DP**	Double-pole
BTM	Bottom	**DRFT**	Draft
C	Celsius	**EA**	Each
CAP	Capacitor	**ECON**	Economizer

10-1

ABBREVIATIONS *(cont.)*

EFF	Efficiency	**GALV**	Galvanized
ELEC	Electric	**GENR**	Generator
ELEM	Element	**GRAV**	Gravity
ELIM	Eliminator	**GRD**	Guard
ELV	Extra-low Voltage	**GRN**	Green
EMER	Emergency	**GRND**	Ground
EMC	Electro-magnetic	**GSKT**	Gasket
	Compatibility	**H**	Transformer,
EMF	Electro-motive Force		primary side
EMI	Electro-magnetic	**HBC**	High Breaking
	Interference		Capacity (fuse)
ENCL	Enclosure	**HDWR**	Hardware
ENER	Energy	**HE**	High Efficiency
EQL	Equalizer	**HLDBK**	Holdback
ES	Energy Saver	**HLF**	Half
EW	Each Way	**HLW**	Hollow
EXP	Expansion, Exposure	**HNDL**	Handle
EXT	Exterior, Extension,	**HNG**	Hinge
	Extrusion, External	**HNGR**	Hanger
F	Fahrenheit, Fan,	**HORZ**	Horizontal
	Frequency	**HOUS**	Housing
FACT	Factory	**HP**	Horsepower, High
FC	Footcandles		Pressure
FEM	Female	**HRC**	High Rupturing
FG	Fiberglass		Capacity (fuse)
FHLDR	Fuseholder	**HTG**	Heating
FILT	Filter	**HV**	High Voltage
FIT	Fitting	**HZ**	Hertz
FLC	Full-load Current	**I**	Symbol for electric
FLRS	Flourescent		current
FM	Factory Mutual	**IC**	Integrated Circuit
FPM	Feet per Minute	**ID**	Indoor, Inside
FR	Fire-rated		Diameter/Dimension
FRCD	Forced	**IEEE**	Institute of Electrical
FRM	Frame		and Electronics
FRSHN	Freshener		Engineers
FRZR	Freezer	**IGN**	Ignition
FS	Float Switch	**IND**	Indicator, Inducer
FTS	Foot Switch	**INI**	Initiation, Injection
FURN	Furnace	**INLN**	Inline
G	Gram	**INNR**	Inner
GA	Gauge	**INSP**	Inspection

ABBREVIATIONS *(cont.)*

INSRT	Insert	**MFFLR**	Muffler
INST	Installation	**MG**	Milligram
INT	Interlock, Interrupter, Internal, Intermediate	**MH**	Man-Hour, Manhole
		MHZ	Megahertz
INTG	Integrated	**MI**	Mineral-insulated
INTK	Intake	**MM**	Millimeter
INTR	Intermittant	**MN**	Main
INTRF	Interface	**MNT**	Mount
INVRTR	Inverter	**MNTR**	Monitor
IR	Insulation resistance	**MOD**	Module, Modulator
J	Joule	**MTR**	Motor
JNCT	Junction	**MTRL**	Material
JNT	Joint	**MV**	Milivolt
K	Thousand, Kelvin	**MW**	Megawatt
KCMIL	One thousand circular mils	**NA**	Not Applicable, Not Available
KG	Kilogram	**N/C**	Normally Closed
KIP	1000 Pounds	**N/O**	Normally Open
KNB	Knob	**NEG**	Negative
KNRL	Knurled	**NEMA**	National Electrical Manufacturers Association
KVA	Kilovolt Ampere		
KYPD	Keypad		
LEV	Level	**NEUT**	Neutralizer
LG	Large, Long, Length	**NLA**	No Longer Available
LH	Left Hand	**NLB**	Non-Load-Bearing
LIQ	Liquid	**NOC**	Not Otherwise Classified
LL	Live Load		
LMT	Limit	**NON**	Non (without)
LNR	Liner	**NP**	Not Protected
LP	Low Pressure	**NUM**	Numerical
LS	Lump Sum	**NZZL**	Nozzle
LTCH	Latch	**OB**	Outboard
LTHR	Leather	**OBS**	Obsolete
LUB	Lubricant	**OD**	Outdoor, Outside Diameter/Dimension
LV	Low Voltage		
LVL	Leveler	**OFST**	Offset
M	Thousand, Material, Motor	**OH**	Overhead
		O & P	Overhead & Profit
MAG	Magnetic	**OPP**	Opposite
MAN	Manual	**OPTM**	Optimizer
MD	Maximum demand	**ORG**	Original
MDFR	Modifier	**OS**	Oversize

ABBREVIATIONS (cont.)

OSHA	Occupational Safety and Health Act,	**REA**	Rural Electrification Administration
OVL	Oval	**RECEPT**	Receptacle
OVRLD	Overload	**REF**	Reference
PD	Potential difference	**REG**	Regulator
PEN	Combined protective and neutral	**REM**	Remote
		REPL	Replace
PERF	Perforated	**RES**	Resilient
PERM	Permanent	**REV**	Reversing
PLG	Plug	**RFI**	Radio frequency interference
PLNM	Plenum		
PLR	Polarized	**RFLC**	Reflector
PLT	Pilot	**RH**	Right Hand
PLY	Pulley	**RLF**	Relief
PMP	Pump	**RLLR**	Roller
PNL	Panel	**RLY**	Relay
PNT	Paint, Point	**RMP**	Ramp
POS	Position	**RMS**	Root-mean-square (effective value)
PPM	Parts per Million		
PRI	Primary	**RMT**	Remote
PRMR	Primer	**RND**	Round
PSF	Pounds per Square Foot	**RPB**	Rapid Press Balance
		RPM	Revolutions per Minute
PSI	Pounds per Square Inch		
		RPR	Repair
PSIG	Pounds per Square Inch Gauge	**RS**	Rolled Steel, Rapid Start
P & T	Pressure and Temperature	**RSR**	Riser
		RSRV	Reserve
PVT	Pivot	**RST**	Reset
PWR	Power	**RSTR**	Restrictor
QTY	Quantity	**RTNR**	Retainer
R	Resistance	**RTR**	Rater
RBBR	Rubber	**RTRN**	Return
RCD	Residual current device	**RUB**	Rubber
		RVT	Rivet
RCPT	Receptacle	**RWRK**	Rework
RCTR	Reactor	**S**	Second - unit of time
RCVR	Receiver, Recovery	**S4S**	Surface 4 Sides
RCYC	Recycling	**SCFM**	Standard Cubic Feet per Minute
RDCR	Reducer		
RDUN	Redundant	**SCH**	Schedule

ABBREVIATIONS (cont.)

SCKT	Socket	**TRANS**	Transistor, Transmission
SCRN	Screen	**TRM**	Terminal
SDDL	Saddle	**TRNSD**	Transducer
SEC	Secondary	**TRSN**	Torsion
SFCA	Square Foot Contact Area	**TURB**	Turbulator
		U	Symbol for voltage (alternative for V)
SHD	Shade	**UL**	Underwriters Laboratory
SHLD	Shield		
SHNT	Shunt	**UNLD**	Unloader
SLNT	Sealant	**UPLS**	Upholstery
SM	Small	**URD**	Underground Residential Distribution
SP	Static Pressure, Single Pole, Self-Propelled		
		UTIL	Utility
SPD	Speed	**UV**	Ultraviolet
SPRT	Support	**VAC**	Vacuum
STAT	Status	**VAR**	Variable
STBL	Stabilizer	**VAV**	Variable Air Volume
STC	Sound Transmission Coefficient	**VEST**	Vestible
		VLV	Valve
STP	Standard Tempressure and Pressure	**VNT**	Vent
		VPR	Vapor
		VS	Variable Speed
STRNR	Strainer	**WE**	White Enamel
STRT	Start	**WF**	Wide Flange
SUCT	Suction	**WHL**	Wheel
SUPP	Supplemental, Supply	**WR**	Water Resistant
		WTR	Water
SVC	Service	**XX-OHM**	Ohm Rating
SW	Switch		
SYS	System		
T	Time		
TC	Tray Cable		
TEMP	Temperature		
TFLN	Teflon		
TGGL	Toggle		
TMPL	Template		
TMPR	Tempering		
TNNL	Tunnel		
TPE	Thermoplastic elastomer		

About The Author

Adam Ding is a professional estimator with extensive experience in a variety of commercial, residential, institutional, industrial, and infrastructure projects. He holds a Master's degree in Building Construction from Auburn University.

Adam has had a successful career in estimating projects of different sizes, ranging from large-scale cost planning to detailed trade take-off. Having bid countless construction jobs, he developed a unique and easy-to-understand estimating approach.

Adam taught computer classes in Auburn University and also owns the copyrights for a number of computer estimating programs. Currently a registered member in the Canadian Institute of Quantity Surveyors (CIQS), he continues to provide cost estimating services to building professionals in North America, Canada, and Asia Pacific regions.